The Hardware OS

Author: Steve Embleton

The operating discipline for hardware programs — how requirements, decisions, and risk stay grounded in what the build shows.

thehardwareos.com

Publication Details

The Hardware OS

Author: Steve Embleton

Publisher: thehardwareos

First edition: 2026

ISBN (eBook): 979-8-9967292-1-0

ISBN (Paperback): 979-8-9967292-0-3

For permissions and inquiries: book@thehardwareos.com

Dedication

To my loving family.

A Note on the Stories

The programs in this book are invented. They are composite scenarios built to make common failure patterns concrete and easy to follow — not accounts of any specific company, project, or employer.

The failure modes are real. Every dysfunction described here has been witnessed repeatedly, across different industries and different eras of hardware development.

The names, companies, and details have not been changed. They were never there to begin with.

The Hardware OS, in one page

What this operating system looks like when it's working

The concrete test is not a metric. It is a set of people doing their jobs without friction.

Your exec can leave Friday afternoon. Not because the program is done, but because the gate status is real, the risk register surfaces what would otherwise surprise them, and every open decision has a named owner and a date. They are not anxious because they are not holding the program truth in their head. The one-page status is holding it for them, and they trust it.

Your PM stops asking "where are we?" A PM who trusts the schedule and the risk log shifts from chasing daily status to clearing real blockers. The artifacts answer the question before it is asked. That is not a nice-to-have — it is the measurable output of Layers 3 and 4 working. A PM who stops pinging is a PM who has something reliable to read.

Your engineering team makes decisions, not presentations. When requirements are managed hypotheses with visible evidence, when gate packets come from live records instead of slide heroics, and when the schedule is built from real dependencies, the team spends its time on the technical problem — not on reconciling contradictions between what the slides say and what the test floor sees.

These are the outcomes the OS is designed to produce. The chapters are the mechanism.

◎

Five operating layers produce these outcomes. Knowing the layer a chapter belongs to tells you what it is *for*, and what it is **not** for.

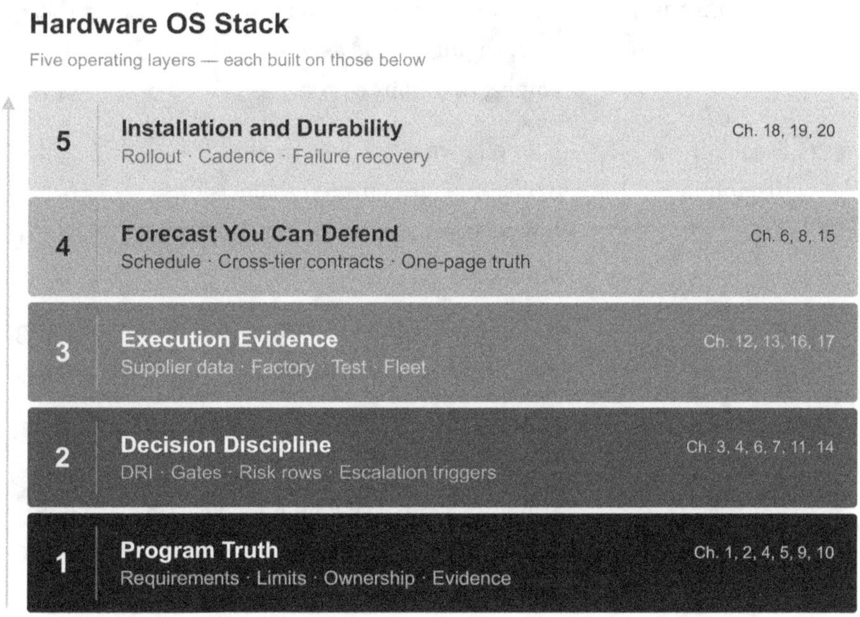

Figure 1: Hardware OS Stack — five layers from Program Truth at the foundation to Installation and Durability at the top

Layer	What this layer guarantees	Ch.
1. Program truth	One source of truth, one ownership rule, requirements as managed values, decisions as records	1, 2, 4, 5, 9, 10
2. Decision discipline	Named owner for every open decision; gates entered as decisions, not reviews; risk rows that can trigger a mitigation action	3, 4, 6, 7, 11, 14
3. Execution evidence	Evidence from the test floor, supplier, and fleet that is traceable to a requirement and current enough to change a decision	12, 13, 16, 17
4. Forecast you can defend	Schedule from real dependencies, tier obligations made visible, executive one-page truth	6, 8, 15
5. Installation and durability	Predictable OS install sequence; cadence that survives personnel changes; named scope limit so the OS does not overpromise	18, 19, 20

A chapter can appear on more than one layer when its mechanism genuinely spans them. Gates show up in decision discipline and in forecast because a gate decision both changes a controlled baseline *and* moves the forecast. Automation at concept (**Ch. 7**) belongs in decision discipline, not execution evidence: it fixes access, presentation, and traceability fields before the line is producing population-scale floor and fleet returns. The layer is not a category; it is a declaration of what the chapter must guarantee for the OS above it to hold.

Reading straight through, the layers assemble in order: truth first, decisions about truth, evidence that supports the decisions, a forecast that holds the evidence and decisions together, then finally the install plan that makes the whole system survive contact with the next program.

How to use this map

Use this page as a map you come back to. Each chapter states which mechanism it introduces; this page says which layer that mechanism lives in. If a chapter feels repetitive, check which layer it is on — it may be introducing the same vocabulary in a different operating context (a different scale, a different tier, a different phase of the program).

If you are in a specific role:

- **Engineering lead:** start with Layer 1 (program truth). Everything else is scaffolding on top of that foundation.
- **PM:** Start with Layer 4 — the schedule, cross-tier contracts, and one-page status are the artifacts a PM reads and defends daily; Layer 2 explains why the gates and risk rows that feed those artifacts are built the way they are.
- **Executive or sponsor:** the one-page truth (Ch. 6), cross-tier contracts (Ch. 15), and schedule integrity (Ch. 8) are the direct

executive-facing chapters. The chapters in Layers 1 and 2 explain why those artifacts hold when the program is under pressure.

- **Hardware founder or technical lead running their own program:** start with Layer 1 — the gap between a technically strong founder and a program that survives the first handoff or scale-up is almost always missing ownership records and requirement locks, not missing technical depth. Layer 5 (Ch. 18–20) names where the OS does not substitute for regulated or specialist obligations.

Part I - Why Hardware Programs Break

1. The Missing Manual

Monday.

6:47 a.m.

Parking-lot light still on — messages already stacking before the standup.

The thermal limit changed again.

Mechanical already cut metal to last week's number. Electrical has a new test trace that says the old number was optimistic. The supplier says they never got the last revision anyway. The program manager is trying to hold the date everyone committed in the first quarter — and your executive wants a one-page update by noon: *Are we still on track or not?*

Everyone is working. Nobody is slacking.

The program is still drifting.

You have watched this movie. It is rarely weak engineers. What is almost always missing is a manual for the process: no agreed way for work to cross team lines, no shared rule for what "locked" means when new data shows up, no honest tie between the schedule and long leads plus what test actually measured, and no single page where that story stays true long enough to survive the next status call.

Process means the written habits: who owns the handoff, where the numbers live that the factory is allowed to build to, who may change that number after a test, and how everyone who needs to know finds out before they order steel. Not paperwork stapled on after the design is "done." The thing that keeps the design from quietly forking inside three different inboxes.

Most engineering leads earn the role on technical judgment, then get judged on skills nobody prints in the promotion packet: who owns the step when mechanical's output becomes electrical's constraint; what has to exist after a meeting so people actually build to the same assumption; what belongs on a one-page status; how a limit moves when evidence says it must; when to pull an executive in before a week turns into a quarter. Nobody hands you a course for that bundle. The industry still behaves as if you learn it by walking past the right cubicles long enough.

That gap costs money the same way bad parts cost money: slips, rework, supplier scrap, trust thinning with the people paying the bills. The same scenes on repeat: no owner on the interface line in writing, executive-facing pages that read confident while build and test already know the open item, suppliers going quiet when the drawing set forks, the stretch of calendar where cheap learning has become expensive metal.

This book closes that gap.

The PM in the room

The PM is usually the easiest face to blame when the review gets loud. They are also the one left holding ship windows someone already showed a paying customer while the technical facts are still moving. When what the team knows and what leadership sees do not match, someone still has to turn what actually happened into a plan someone can fund. Facts in hallway fragments and slide footnotes mean the PM keeps asking "where are we?" while mechanical, electrical, and supply each carry a different piece of the answer.

This is not a personality failure. Missing process: named owner for the handoff, one-page story that matches build and test, written decision when a tradeoff changes, early pull on risk and long leads while you can still steer. When the week stops burning on the same unanswered question, PMs get room to clear blockers instead of running the same worry past everyone again. Those habits start with leads, because that is where they get installed first or not at all.

From the outside it can look like temperament. It is almost always facts that never landed in one place. Back to the list you can actually fix.

What "missing manual" looks like

You see the same failure modes often enough to stop calling them bad luck.

Handoffs nobody owns. Interfaces, limits, and supplier cut-ins sit between org charts. Everyone assumes someone checked. Nobody can point to the sheet that is supposed to be true tomorrow morning.

Paper locked, reality unlocked. Numbers and dates live in mail threads or decks, not where build, test, and supply chain actually run.

Heroic individuals, fragile program. The work gets done because one person holds the context — who owns the call, where the open items live, what the real number is. When that knowledge is not in a record, work that depends on it cannot close. People push in parallel on the same question, each carrying a different piece of the answer, none of it forced into a single decision.

Late truth. Risk and dependency never land in writing where PM and executives can help early, so the surprise shows up first in an exec review, then in metal, then in the field. Wrong altitude for cheap fixes.

Schedule theater: the physics, procurement, and test facts that drive the date moved weeks ago, but the shared plan never caught up. The meeting defends loyalty to a date instead of updating the plan to match the facts. Name it that way and you can at least stop adding your own layer to it.

Field test: pick any open technical decision on your current program. Ask who owns a change to it and where the current number lives. If those answers diverge across three people, or if no one can point to a written record, the program is running on memory — and the manual is missing.

The standard response when a program slips is to add a tool. Sprints. Scrum. Jama for requirements. Jira for everything — more Jira, always more Jira. The boards get built. The training happens. Six months later the same program is behind the same way. The tool was not the problem and cannot be the solution. Jira cannot tell you who owns the handoff when no one does. A sprint ritual cannot lock a requirement that was never locked. The failure is upstream of the software: unowned decisions, numbers nobody committed to, schedules built on optimism rather than lead times. None of that requires new software to fix. It requires writing the right things down and naming who owns them.

None of those get fixed by working harder on the same scattered pattern. I came to this late. A few different employers showed me what I had not been naming.

Three programs, one lesson

Three stretches of work. Same lesson as the list above.

Same people, different written habits. What changed was control discipline: one controlled source for the build limit instead of four slide versions, a named signer and date for post-test changes, and direct reporting from the engineer who did the work instead of each layer rewriting the message upward.

Early in my career I saw programs where timelines slid again and again with very little forcing function: no single story of one plan, one date, one record people actually used. People were busy. Key decisions stayed open anyway.

Later I touched energy hardware where right-to-left schedule pressure was not a slide. Survival. Long leads and thermal work landed in the same month as procurement. PMs there were doing real work keeping motion visible while engineering learned what had to be surfaced and written down before it burned the path. Risk and dependency still hurt when they arrived late. Expensive in time and metal.

The strongest program rhythm I have seen personally was on a cell program: tight requirements, real phase gates, a PM who made state obvious to everyone, engineers who owned critical work in front of leadership without each manager filtering and re-editing the message on the way up. Strong process, and the program still needed a technical lead to keep the team from circling the same narrow pocket in the design space for a quarter. Good process did not erase leadership. It focused it.

More recently, in larger power-electronics organizations, the biggest wins I have seen are rarely a clever topology tweak. They look more like pulling a program out of opaque reporting, soft requirements, and timelines that slip without anyone writing down why. The biggest gain is often process — engineering cost and time recovered — not a topology or architecture win.

Those sketches point at the same conclusion as the opening: nobody is trying to fail. The operating manual for how decisions and numbers move is what is missing.

What this OS covers — and what it does not

Operating habits and records for hardware programs: ownership, spec integrity, gates that mean something, risk you can read, fleet and factory learning used as instruments, automation choices made early enough to matter, schedules tied to real dependencies, supplier truth that matches what build and test see.

Primary reader: the engineering lead, the person who ends up owning integration whether or not the title caught up. PMs and executives too. Same facts in the same order the lead needs, without inventing another standing meeting to get them.

Not a textbook on formal drawing rules, structured multi-factor experiments, or simulation theory. Those tools matter; pull specialists when the program needs them. You get when to insist on that class of work, what "good enough to decide" looks like, how the result feeds a gate and a one-page status.

Not a replacement for design history, functional safety, or regulatory paperwork. The final chapter names the boundary so nobody confuses these habits with compliance.

Process is the product — the system being built while the hardware is being built. The question that follows is how to run that without it becoming religion.

2. Process Is the Product

Monday morning. Two programs had just failed the same end-of-line leak test.

On the test stand, the two failures looked identical.

Both teams have competent engineers, a supplier in the loop, and a date that will not move without consequences. Both have test data that says the current build is not stable enough to ship.

One program contains the issue in a day and restarts with a revised control plan. The other burns two weeks arguing whether the failure is fixture drift, operator variation, or the wrong revision on the floor.

The deciding variable is decision plumbing: how test evidence becomes a signed plan.

Same physics, different outcomes

Program A has one controlled source for the leak-rate limit and fixture calibration state, with revision history and owner. When failures spike, the containment decision, root-cause owner, and restart criteria are written in one decision record with signer and date. The PM one-page status pulls from that same record, so leadership, manufacturing, and supplier all see the same state. The date moves, but it moves once and with reasons.

Program B has the drawing revision in one system, the work instruction in another, and supplier evidence in email. Manufacturing says fixture drift. Design says tolerance stack. Quality says operator variation. The status deck still shows last week's confidence color because nobody owns the merge from evidence into decision. The PM spends the week reconciling contradictions instead of clearing blockers. By the time leadership sees the real risk, cheap fixes are gone.

Same technical problem. Different operating system.

Teams call this "execution," but the mechanism is simpler: the path from evidence to decision is either designed or improvised. If it is improvised, each shock costs more than it should.

What process actually is

In this book, process is not ceremony. It is the minimum written system that keeps one program reality across functions.

At minimum, that system has:

1. **Named ownership** for decisions and handoffs.
2. **One source of truth** for active limits, requirements, and revisions.
3. **Decision records** that explain what changed, why, and who approved it.
4. **Gate criteria** that decide, not just review.
5. **One-page status** that matches the underlying record.

6. **Escalation rules** for when evidence breaks the current plan.

If one of those is missing, the team does not fail immediately. It fails noisily over time: duplicate work, delayed truth, meetings that replay the same argument, and last-minute surprises that look like "bad luck."

This is why process belongs in design work, not in a PM appendix. It changes what gets built, when it gets built, and what can be defended to customers and leadership.

Why weak process makes technical problems expensive

Weak process adds cost in four ways.

First, it creates **latency**. Good data exists, but it takes days to become a decision because nobody owns the transition from test result to plan update.

Second, it creates **rework**. Different groups act on different versions of the truth, then spend time undoing each other.

Third, it creates **risk blindness**. The one-page status reports confidence while the underlying record reports uncertainty, so management acts on stale optimism.

Fourth, it creates **trust debt**. Suppliers, PMs, and executives stop trusting the current number because the number keeps changing without a visible decision trail.

None of that is abstract. It shows up as scrap, missed windows, overtime, and avoidable churn in people who are already stretched.

THE OPERATING RULE

Treat process artifacts (written program records) as product artifacts.

If a limit, dependency, or risk can change build, cost, or date, it needs an owner, a controlled record, and a visible decision path. If it has no owner or no record, treat it as not real yet.

This rule is strict on purpose. Hardware programs punish ambiguity late. Remove ambiguity when evidence first appears, before purchasing and build decisions fan out.

A corollary worth keeping visible: every artifact in this system should reduce latency, rework, or ambiguity. If a record, checklist, or status field does not demonstrably do one of those three things, retire it. The minimum system listed above is exactly that — minimum. Do not add to it until the minimum proves it reduces argument and lag on your program.

WHAT CHANGES THIS WEEK

If your program is already moving, do this before adding new templates:

1. Pick one active cross-team limit that is causing repeated meetings.
2. Name one owner for that limit.
3. Declare one controlled location as the live value.
4. Write one short change log entry format: what changed, why, who approved, impact.
5. Make the next one-page status pull from that record.

That is enough to expose whether your current process is implemented or cosmetic.

Field test: pick the highest-stakes open limit on your current program. Can you name its owner and point to a controlled written record of its current value? If the answer to either question is no, the process is cosmetic at that limit — regardless of what the status deck says.

Install one narrow control loop first. Prove it reduces argument and lag, then extend.

Install one owner-record-decision loop on a single stuck issue this week and watch what surfaces first. A missing owner, a number nobody controls, a decision no one will sign — each is a different failure with a different fix, and naming which one you are actually fighting is the whole job.

3. Failure Patterns: Chaos, Drift, and Unowned Decisions

The minimum control loop is running. What it cannot do on its own is tell you which failure pattern is breaking it — or where to look first.

By Day 19 of the handheld-controller EVT, the leakage-limit decision was in its third meeting and a $42k fixture order sat on hold, slipping with every meeting that failed to close it.

Test said one thing. The status page said another. Procurement had already acted on a third number, because nobody had marked which one was live.

Three people in the room had calendar proof they had worked the issue. The program had no record of who owned the live number or when it was last confirmed — so the decision window closed again, the same way it had twice before.

This is where teams waste months: treating recurring operating failures as isolated incidents.

Why naming patterns matters

Most postmortems describe events. Useful postmortems name patterns.

An event is "supplier lot failed incoming check." A pattern is "we had no owner for the handoff from incoming data to build decision, so three teams acted on different states for six days."

If you only name events, you fix the last fire. If you name patterns, you remove a class of fires.

Three patterns show up together in hardware programs:

1. **Chaos**: too many truths active at once.
2. **Drift**: reality moved, but plan/status did not.
3. **Unowned decisions**: the question is known, but no person is accountable for closing it.

They look separate in meetings. In practice, closure slips spread parallel assumptions and stale status, so the three patterns reinforce each other.

Pattern 1: Chaos

Chaos means multiple conflicting versions of the same fact are active at once.

Symptoms are easy to spot:

- Two "latest" revisions in circulation.
- Lab evidence in one channel, supplier evidence in another, decision in neither.
- A confidence-green status page built from stale assumptions.

Chaos cost is immediate: rework, mis-buys, and wasted meeting cycles spent reconciling state before real problem-solving even starts.

The core mechanism is missing control of truth state. If nobody can answer "which value is live, where, and since when," the team is in chaos even when people feel organized.

Pattern 2: Drift

Drift starts after a legitimate change.

New evidence arrives. A requirement shifts. A supplier lead time stretches. Nothing dramatic happens that day. One week later the schedule, one-page status, and downstream tasks still assume the old world.

That lag is drift.

Drift cost compounds quietly:

- Date risk is discovered late.
- The risk register shows lower risk than the evidence supports, because nobody updated it after the change.
- Teams optimize for targets that no longer match physics or supply reality.

The core mechanism is slow or missing propagation of change. A change without a written, owned update path is not a change. It is a pending surprise.

Pattern 3: Unowned decisions

Unowned decisions usually sound like "we keep discussing this."

The question is clear. The data is mostly available. The trade is known. But there is no named person responsible for closing the decision, documenting it, and updating dependent work.

So the decision becomes a recurring meeting topic instead of a completed program action.

Unowned decisions create both chaos and drift:

- While closure is delayed, multiple interim assumptions spread (chaos).
- As closure slips, the plan and status diverge from current evidence (drift).

This is why ownership is not a management preference. It is a technical control.

How the three patterns chain together

They usually run in this order:

1. A decision is unowned, so closure slips.
2. During the slip, teams run different assumptions (chaos).
3. Status and schedule lag behind reality (drift).
4. The eventual correction is expensive because many downstream actions already committed.

You can interrupt the chain at any step, but the cheapest break point is first: assign ownership and close the decision while the blast radius is still small.

Fast diagnostic you can run this week

Pick one active cross-functional issue and ask five questions:

1. What exact question must be decided?
2. Who is the owner of closing it?
3. Where is the current agreed value or decision recorded — the one the team is building and testing to right now?
4. What changed in the last week, and where is that recorded?
5. Which downstream artifacts were updated after the change?

If your team cannot answer all five in under ten minutes, you are not dealing with one bad meeting. You are inside one or more of these patterns.

What to fix first

Do not try to solve chaos, drift, and ownership with three separate initiatives.

Field test: name the most-contested open decision on your current program. Can you point to one controlled record with a current value, one named owner, and a confirmed date? If not, you have an unowned decision — and the recurring meeting is the symptom, not the problem.

Day 19 of the handheld-controller EVT — the $42k fixture order on hold, the third meeting on the same leakage-limit decision — resolved when one owner was named for the live value. A single field was added to the existing status record: current live value, owner, and date last confirmed. The next week's review ran the five-question check in seven minutes. The fixture order cleared. The decision did not come back to a meeting.

Part II - The Hardware OS Core

4. DRI Ownership and Decision Authority

Three failure patterns run together in hardware programs: chaos, drift, and unowned decisions. Unowned decisions are the first break point because without a named person accountable for closure, the minimum control loop has no one to run it.

The decision record read: "Thermal limit — discussed, will revisit." Owner: blank. Date last confirmed: blank.

That was the whole entry, late on a Monday, with the question clear and the data good enough. The meeting still had not decided, because nothing on the record said who was allowed to.

By the next week the same topic was back in three meetings, each working from a different "current" assumption.

This is not a communication problem. It is an ownership design problem.

DRI means closure, not coordination

A Directly Responsible Individual (DRI) is not "the person who schedules the meeting."

A DRI is the person accountable for:

1. framing the decision question,
2. collecting required inputs,
3. proposing a decision by date,
4. recording the outcome and rationale,
5. pushing updates into dependent artifacts (the controlled records the decision touches).

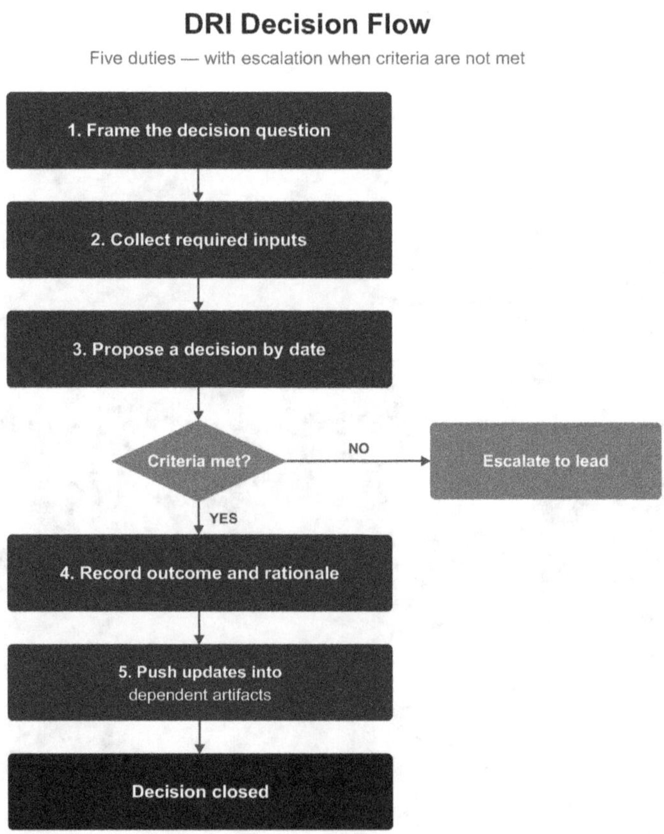

Figure 2: DRI Decision Flow — five duties with an explicit escalation branch when criteria are not met

If those five are not explicit, you do not have a DRI. You have a coordinator.

The lead does not need to personally own every DRI role. Naming a DRI is itself a lead responsibility. An engineering lead who surfaces an electromagnetic interference (EMI) concern can frame the decision question – what does this component need to meet, against what test condition, by what date — and assign that question to another engineer. That engineer becomes the DRI: they own defining the requirements, assessing the risk, setting next actions, and driving to a date.

What the lead does not do after that assignment is re-own the path. In reviews, the lead's questions are limited to three: is the closure criterion met, do you need help escalating with another team, and does the one-page status need updating? The lead does not have enough context to judge whether the work is at risk, whether a due date is realistic, or whether escalation is warranted. Those calls belong to the DRI. The lead relies on the owner to surface them.

Authority must match responsibility

The most common ownership failure is giving responsibility without authority.

"You own this thread" means nothing if the owner cannot:

- call the decision when criteria are met,
- escalate when criteria are blocked,
- and enforce update of the one-page status and dependent records.

Authority does not mean unilateral control over everything. It means a known decision boundary and a known escalation boundary.

A battery-pack team hit this exact gap during a stretch where several problems were running in parallel: cell components were arriving out of expected formation tolerance, one material was on allocation, and the laser welding process was generating consistent leak-test failures that nobody could pin to a single root cause. Three functions — process engineering, materials, and the cell supplier — each had a stated position, but no one had a named DRI or a decision date. For ten days, procurement held orders to two different material specifications while the debate continued. The weld failures in particular were resisting quick resolution; each team's proposed fix was blocked or qualified by another team's constraint.

The fix was not a task-force. It was a one-line revision to the ownership record: one process engineer was named DRI for the weld-process decision with explicit authority to call the hold on any build lot when leak-test failure rate exceeded a stated threshold — without waiting for cross-function consensus. The authority boundary was written, not assumed. Once it was, the process engineer called the hold, drove the weld parameter validation without waiting for permission, and closed the leak issue in the next build cycle. The earlier delay was not an analysis problem. It was the absence of a named person who could call the hold.

What breaks when ownership is fuzzy

Shared ownership sounds collaborative and often behaves like delay.

Failure signatures:

- Decisions reopen because nobody is accountable for final closure.
- Teams hold parallel assumptions while waiting for "alignment."

- Escalations happen late because no one is responsible for raising the flag.
- PM and exec updates become negotiation artifacts instead of state artifacts.

The same gap shows up quieter. A bus-bar gauge decision on one program was "made" three separate times over eleven days — each time a different function reopened it in a different meeting, because no record named who owned the close or what would end the debate. The status slide stayed green the whole time; the real state was one decision being relitigated on a loop. What ended it was not more analysis but a single line naming the owner and the date past which the call would stand.

The cost is not just time. It is trust in the decision system.

Minimum ownership map for Week 1

You do not need an org redesign to fix this. You need a narrow map:

1. Top 10 active cross-functional decisions.
2. One DRI per decision.
3. Decision deadline per decision.
4. Required inputs per decision.
5. Decision authority level (team, lead, exec).
6. Escalation path if blocked.

Publish this map where engineering and PM both work from it. If it lives only in one lead's head, it is already stale.

Decision record standard (minimum)

Every closed decision needs one short record:

- decision question,
- decision made,

- alternatives considered,
- why this choice now,
- owner/sign-off/date,
- downstream artifacts updated.

No long memo required. One durable record is enough to stop re-litigation.

The decision record can live in any format — a plain text file, a shared document, a wiki page — as long as it has three properties: a static link anyone on the program can reach today and six months from now, a date stamp or edit history that shows when it was last changed, and a location that is not buried in a conversation thread.

What disqualifies a tool is not its name but its behavior. Chat and email fail as decision records because state drifts — the decision is somewhere in a scroll of messages with no stable address, no clear owner field, and no way to confirm which version is current. A decision sent in a chat message is invisible to anyone who joined the thread late and unrecoverable once the channel is archived.

The question to ask of any tool: can I send someone a direct link to this record right now, and will that link still resolve in a year? If the answer is no, that tool is not suitable for decision records regardless of what else it is used for.

The fields that must be present — owner, decision question, evidence, rationale, affected artifacts, approver, date — are the requirement. The container is implementation detail.

Decision record: bad vs corrected

Field	Bad (incomplete)	Good (decision-grade)
Decision	Revise thermal limit	Revise thermal limit from 75 °C to 78 °C per chamber measurement
Owner	TBD	Thermal lead
Rationale	*(blank)*	Chamber measured 78 °C steady-state; boundary condition error found in model; contact-resistance verified by direct measurement
Downstream updates	*(blank)*	Supplier package flagged for re-qual, thermal risk row updated, gate one-page revised
Approver	*(blank)*	Program lead
Date	*(blank)*	Day 9 post-chamber run

Escalation without drama

Escalation is a design feature, not a social failure: it fires when trigger conditions are hit, not when room politics allow it.

Define escalation triggers in advance:

- missing critical input by date,
- unresolved cross-function conflict,

- risk threshold exceeded,
- customer/compliance impact.

When trigger conditions are objective, escalation is faster and less political.

If you can name the DRI for every open program decision today, and each DRI has proposed a decision by a date, the OS is doing its job. If any question is 'TBD / team / committee,' it is not.

Decision classes: not every call needs the same overhead

Not every decision warrants the full DRI machinery. The overhead should match the reversibility.

A **reversible decision** — component selection before tooling, test parameter adjustment during Engineering Verification Testing (EVT) — can move fast with a lighter record and a lower authority threshold. The DRI still owns closure, but speed is the right bias. The cost of being wrong is small; the cost of being slow is real.

An **irreversible decision** — tooling release, supplier qualification, customer specification commitment — requires the full map: written authority boundary, decision record, downstream artifact updates. Replaying an irreversible decision is expensive enough to justify the added friction before calling it.

A **build-to-learn unit** is a third class. The acceptance criterion is "what did we learn?" not "did it pass the requirement?" Treating a build-to-learn result as a production failure is a scope error. Name the unit's class before it hits the floor. The DRI for a build-to-learn unit owns the learning extraction, not the conformance call.

If you are not sure which class a decision falls in, ask one question: if we call this wrong, can we undo it before purchasing and build decisions fan out? If yes, it is reversible. If no, treat it as irreversible until proven otherwise.

Field test: pick the highest-stakes open decision on your current program. Who is the DRI? If three people answer that question differently, or if no one will say the name without hedging, the decision has no owner — and the program is waiting for someone else to close it.

5. Spec Integrity and Requirement Hygiene

Ownership defines who closes decisions. What those decisions close on is the requirement record — and a decision is only as good as the requirement it references.

The released drawing package had Rev C on the cover and Rev B in the work instructions — both inside the same folder stamped "released," on a Tuesday morning with the line about to build.

Manufacturing asked which revision was real. Test had validated Rev C assumptions. Procurement had placed orders against Rev B.

The package was "released." The program still had three definitions of done.

What spec integrity means

Spec integrity means every team can answer the same four questions with the same source:

1. What is the current requirement?
2. Why is it that value?
3. What evidence supports it?
4. Who approved the latest change?

If answers differ by function, the program does not have a spec. It has parallel interpretations.

Hygiene failures that look small and cost big

Most failures are not dramatic. They are quiet mismatches.

Rev C.2 changed the leak-test dwell value, but manufacturing work instructions still referenced Rev C.1 while test benches used the new value. That one mismatch forced three days of teardown and retest. The team then found the dwell requirement had no owner and no rationale trail for the original dwell change. Once an owner was assigned and a controlled revision record was created with the rationale, the next revision — a ten-second dwell adjustment — was processed in four hours and reached all affected work instructions the same day.

Requirement statement quality (minimum bar)

A requirement is usable when it is:

- specific (single unambiguous statement),
- measurable (clear units and method),

- bounded (conditions/scope explicit),
- traceable (stable unique ID; named origin),
- owned (named person accountable for quality).

Traceable means the requirement has an identifier that does not change when the value changes — downstream artifacts reference the ID, not the number — and a named origin: the customer spec, test result, or engineering decision that justifies why this requirement exists.

"System shall be robust" is not a requirement. "Leak rate shall be <= X under conditions Y and Z, measured by method M" is.

Revision discipline

Revision control is where integrity usually breaks.

The minimum: one controlled source — same location, same revision label — that every function reads from. If the value lives in a chat message or a personal folder, it is not controlled.

The rest adds durability:

- Change log entry with rationale and evidence link.
- Explicit impact note: which tests, supplier packets, and schedule rows must update.
- Named approver matched to change scope — adjusting a tolerance within a validated range needs the requirement owner's sign-off; redefining what a requirement measures needs broader authority because every artifact built against it is affected.

On one program, a stress analyst revised a structural load limit after a test result, notified manufacturing verbally, and considered the revision complete. The structural load requirement remained at the old value in the controlled record. Manufacturing built to the revised limit; the supplier qualification test ran against the original.

The gap surfaced at first-article acceptance: a three-week hold while the record, the build, and the supplier notification were reconciled.

If a value changes without the controlled source updated, treat the change as unofficial until cleaned up.

Requirement hygiene checklist for active programs

Run this weekly on active high-risk requirements:
☐ Owner named
☐ Current value and revision confirmed
☐ Source/evidence linked
☐ Acceptance method defined
☐ Last change rationale captured
☐ Downstream artifacts updated (test plan, one-page status, supplier packet)
This is not paperwork for paperwork. It is a control loop for decision quality.

Three implementations that work:

- A spreadsheet with a named-owner column, a current-value column, an evidence column, and a change-log tab — sufficient for most programs and requires nothing to set up.
- A shared document with a fixed requirements table and a change history section at the bottom — works when the team already has a shared document location everyone reads from.
- A version-controlled plain-text file in the same repository as design files — works for small engineering-led teams where file history serves as the revision record.

All three share the same requirement: six fields in one stable location — owner, current value, evidence link, acceptance method, change rationale, and downstream-update record — and a history that shows what changed and when. What disqualifies a tool is behavior: if the current value is in a message thread and the approval is in an email, there is no controlled requirement. There is only a number floating in conversations with no way to confirm which version is live.

Why requirement quality comes before gate design

Gates cannot be strong if requirement quality is weak.

A gate asks, "Are we ready to commit?" If requirement truth is ambiguous, the gate is a well-prepared presentation over hollow data — the review happens, the questions get answered, and nothing said verbally is reliable.

Spec integrity is the substrate under every later control: risk scoring, one-page truth, supplier alignment, and schedule realism.

Field test: If a requirement in your current baseline has no evidence link, no named acceptance method, and no revision history, it is not a requirement. It is a wish with a number. Find three of those today and fix them.

6. Gates, Risk Registers, and One-Page Truth

Requirement quality is the floor. Gates, risk rows, and one-page status are the synchronization mechanism that keeps requirements, decisions, and evidence in one shared view across all functions.

“The deck is clean. We’re green to proceed.” Tuesday mid-morning, gate slides up, and that was the entire case for passing.

No one in the room could answer the question that actually decides a gate: what changed since the last one that alters risk to launch.

A clean deck and a program that can show in two minutes what moved are not the same thing. Only one of them is making a decision; the other is performing confidence.

Gate reviews: decision points, not slide events

A gate is useful only if it closes explicit decisions.

Minimum output of every gate:

1. decision taken (go/hold/re-scope),
2. owner for each follow-up action,
3. date and criteria for closure,
4. artifact delta list — what controlled records changed as a result.

The artifact delta list is the implementation trail: thermal limit revised from 75 °C to 78 °C | supplier re-qual opened | thermal risk row closed | schedule baseline updated to rev 4. Without it, a decision taken in the meeting has no verified follow-through.

No decision, no gate. Just a meeting.

Not every gate is a decision gate. A **readiness review** confirms that planned next-phase inputs are in place — no controlled value needs to change. An **evidence review** surfaces test or analysis results without forcing a decision yet — useful when the data is new and the team needs to absorb it before proposing a change. An **alignment meeting** reconciles cross-tier expectations without triggering a baseline change. A **regulatory or customer staging gate** moves a formal record forward in a controlled lifecycle. All of these have legitimate jobs. The rule "no decision, no gate" applies specifically to events on the program plan named "gate" that are expected to change a controlled baseline, requirement, or release — when those events happen but no controlled value changes, the gate is mislabeled. Rename it accurately: call it a review, write down what it produces, and stop treating it as a gate.

RISK REGISTER: FROM DECORATIVE LIST TO CONTROL TOOL

Most risk registers fail because they track nouns, not decisions.

A usable risk row needs:

- risk statement,
- risk category (schedule / technical / regulatory — name it explicitly),
- owner,
- likelihood/impact basis,
- trigger signal,

A trigger signal names the specific observable condition that forces an action. Example: "any chamber run exceeding 74 °C" — if that reading appears, the thermal risk row escalates to a gate hold regardless of where the schedule stands. Without a named trigger, a risk row stays yellow indefinitely — acknowledged but never forcing action.

- mitigation action,
- decision date,
- current status with evidence link.

If "owner" is a team name and "status" is a color, the register will look active and behave inert.

Risk is not monolithic. A schedule risk — supplier lead-time uncertainty, tariff exposure, a dependency that has not been confirmed — moves through procurement decisions, buffer management, and contract terms. A technical risk — a thermal margin built on a back-of-envelope assumption, a structural calculation awaiting simulation, a supplier process not yet characterized at production rate — moves through testing, simulation, and supplier qualification. The mitigation actions for each are different, the owners are different, and the trigger signals are different. A register that collapses both under a single color produces the worst outcome: owners who are unclear on what kind of action will actually reduce the risk, and leadership reading one signal that covers two completely different problem types.

Each mitigation action should be specific enough that completing it demonstrably reduces the stated risk. When the action closes, the row carries the evidence — test data, simulation result, supplier Cpk, confirmed lead time — not just a status color flip. That evidence is what makes the risk register a decision artifact rather than a status decoration.

RISK REGISTER: BAD ROW VS DECISION-GRADE ROW

Field	Bad row	Decision-grade row
Statement	Thermal	Thermal margin insufficient at full-duty 40 °C ambient
Category	*(blank)*	Technical — thermal margin pending supplier re-qual
Owner	Engineering team	Thermal lead
Likelihood / Impact	Yellow	Likelihood: medium (chamber showed 78 °C vs 75 °C limit) / Impact: high (gate hold, supplier re-qual)
Trigger	*(blank)*	Any chamber run exceeding 74 °C
Mitigation	In progress	Revise thermal limit, re-qual supplier package
Decision date	*(blank)*	Named gate date
Evidence	*(blank)*	Chamber run, contact-resistance measurement

One-page truth: executive view tied to source records

The one-page status is not a summary slide. It is the control surface for shared truth.

Minimum one-page sections:

1. current state vs target (what moved this week),
2. top risks and triggers,
3. decision log deltas,
4. critical dependencies and slips,
5. asks/escalations with owners.

Every line should trace to a source artifact. If leadership cannot drill down from line to record, confidence will exceed reality.

A one-page status that traces every line to a live record is the artifact leadership can act on without a phone call. Not because the program has no risk — every hardware program has risk — but because the risk is named, owned, and visible in the register. The exec reads the register instead of calling, because the register surfaces what would otherwise surprise them. An engineering lead who has built a traceable one-page is an engineering lead whose exec has something reliable to read.

The one-page can live in any shared location that supports a stable link and a named owner: a wiki page, a version-controlled doc, a live dashboard, a weekly-refreshed slide. The format does not determine quality. What determines quality is whether each line links to a live source record — test result, risk row, decision entry, schedule baseline — and whether a reader can follow that link to verify the claim without asking anyone.

One-page status: bad line vs traceable line

Bad: > Thermal: On track — team working the issue

Good: > Thermal: limit revised to 78 °C per chamber | thermal risk row closed at gate | supplier re-qual in progress | Cold-plate extrusion: 11-day slip, recovery in schedule rev 4

Update rhythm that keeps it honest

Cadence matters more than formatting.

Recommended rhythm:

- risk register refreshed at least weekly by owners,
- one-page refreshed on fixed cadence before leadership touchpoints,
- gate packets pulled from live records, not rebuilt from memory.

Late updates are not neutral. They create drift by definition.

Failure mode: deck replaces record

A deck works fine as the one-page status surface if it is updated by its owners on a fixed cadence, links to source records rather than restating them, lives at a permanent shared link, and is accessible to every tier. The failure mode is not the format.

When any artifact — deck, wiki, dashboard, or shared doc — stops holding the current state and becomes the thing teams optimize for narrative coherence:

- decisions disappear after the meeting,
- risk ownership blurs,
- status reflects what the team agreed to report, not what the records show,
- and escalations arrive as surprises.

The fix is not a different format. The fix is that whatever format carries the one-page must trace every line to a live record — and the tool holding that record must not be ephemeral. Specs need revision control so two people cannot be working from different versions of the same requirement. The one-page needs a stable, shared link so no one has to ask where it is. Data citations belong in a traceable location — a linked ticket, a controlled test record, a versioned log — not in a chat thread or an email chain. If the team has to reconstruct a decision by searching message history, the information was stored in the wrong place.

Practical install in two weeks

Week 1:

- standardize risk row format,
- assign owner to every active high-impact risk,
- define trigger rules for top 10 risks.

Week 2:

- rebuild one-page from risk + decision records,
- run next gate using artifact delta list,
- reject items that cannot trace to source.

Before Gate G3, the thermal risk row listed "supplier CpK pending" with no decision date, and the one-page line still read "launch confidence: green." The decision had been deferred through two prior gate cycles — each cycle the row stayed yellow and no action date was set. After the gate decision to hold the launch lot and fund gauge redesign, the row gained an owner and a decision date and the one-page delta changed to "launch at risk until CpK ≥ 1.33 evidence dated 2026-05-12." The decision, once it carried a named decision date, closed in one meeting. Gauge data collected two weeks later confirmed the launch lot would have produced a 12% first-pass yield miss at the customer — a containment action that would have cost more than the gate hold ever did. The whole gap was one missing field: a decision date the row had been allowed to live without through two gate cycles.

Status is not forecast

Gates and one-page truth tell you what is true now. They cannot tell you when it will be done. Whether the evidence they need at the next gate will exist — whether the test coverage was designed to produce it — was decided before this gate opened.

Field test: look at your last gate output. Can you name the DRI for each open risk row, point to the evidence that moved it to its current status, and trace every line on your one-page status to a live record? If not, the gate was a meeting, not a decision.

7. Automation Decisions at Concept Stage

Tuesday, 2:20 p.m., concept review wrapped an hour ago, the enclosure geometry locked and the probe-access question never raised.

The design review celebrates fit, function, and cost.

Six months later, the line cannot probe one critical node without custom fixture workarounds that add 40 seconds per unit and destabilize yield.

Nothing is "wrong" with the product design. The automation path was never designed into it.

The automation decision belongs at concept stage, where the per-unit test cost can still change the design.

The core mistake

Teams treat automation as a downstream implementation problem.

It is not. Automation viability is set by concept-stage choices:

- access geometry,
- datum strategy,
- tolerance and stack assumptions,
- test point location,
- part presentation.

Delay those decisions and the program pays in cycle time, scrap, and heroic workaround engineering.

On one actuator program — at concept stage, before geometry was locked — the concept team chose recessed test-pad geometry to save enclosure space. At ramp, that choice forced a manual probing station between ICT and final test and added 52 seconds per unit. At 10,000 units per month, 52 seconds per unit added 144 hours of test-floor time; when the team attempted to automate after tooling was fixed, fixture redesign consumed six additional weeks.

That account is the mechanism in miniature: per-unit seconds scaled to monthly floor demand — arithmetic to own at concept while geometry and tooling still trade, not after ramp turns it into a permanent line tax. On the following program, the same pad-access conflict was caught in week two of concept and resolved with a geometry adjustment before tooling froze.

Concept-stage questions that prevent late pain

Before concept closes, settle targets on:

1. What critical characteristics must be measured in production?
2. How will each be accessed physically?
3. Throughput envelope — order-of-magnitude cycle-time / rate that the line family must sustain (fine station-cycle budgeting

comes after functioning prototype builds, before a test production run).

4. What data must be captured per unit for release and field learning?
5. Which steps are manual at launch and automatable later?

If these targets are unset or fuzzy, automation is already a schedule and yield risk.

Minimum design-to-automation handoff

You do not need full line design at concept stage.

You do need a minimum handoff package:

- current concept geometry and tolerance assumptions,
- critical-characteristics list,
- preliminary test strategy,
- target cycle-time envelope,
- traceability/data requirements,
- open risks and decision dates.

This lets automation, manufacturing, and test challenge assumptions while change is still cheap.

Failure signature of late automation

Late automation programs show the same signs:

- fixtures become product redesign proxies,
- cycle-time misses are "discovered" at ramp,
- manual rework stations become permanent,
- data needed for root-cause is unavailable or too noisy.

The debt is technical and usually gets masked as "we need to keep ramp moving."

What to do if you are already late

If concept is over, you still have levers:

1. Identify top three throughput/yield constraints driven by design-for-automation misses.
2. Classify each as design change, fixture change, or process workaround.
3. Quantify cost of each workaround in cycle time and yield impact.
4. Escalate one design change if workaround cost exceeds threshold.
5. Capture lessons as mandatory inputs for the next concept phase.

Use triage outputs to decide **this** program — design changes, fixtures, workarounds, and what ships. Treat the systemic lessons from that work as binding inputs for **the next concept** on the next program (or major platform refresh), so the same class of geometry miss is expensive once across the portfolio, not twice.

Automation is a decision-quality topic

Automation arguments are often miscast as tooling brand or robotics preference.

The sharper failure mode is sequencing: "automation last" pushes someone toward line and casting-level commitments before product geometry, packaging, presentation, conveyance, fixture points, or tolerance envelopes are settled — then the conversation is improvised and brittle. Decide automation when those edges are treated as known inputs — e.g., can this geometry be self-centered and probed reliably on this datum — and the residual question is narrower.

Under that sequencing failure, what looks like a tooling fight is usually requirement and ownership failure:

- unclear measurable targets,
- late closure on critical characteristics,
- no owner for integrating design/test/manufacturing constraints.

Treat automation as part of the operating system and it becomes predictable. Treat it as "integration detail" and it becomes late-stage volatility.

Field test: If the test coverage plan for your current program includes unit counts, cost per unit, and a pass/fail threshold for automation — and that calculation was done before the concept was locked — the decision was made at the right time. If the test coverage is still 'TBD at D-phase,' the decision is already late.

8. Schedule from Real Dependencies and Honest Critical Path

The status slide shows a launch date that has not moved in three months.

The dependency file under it moved every week.

By the end of Tuesday the two had drifted far enough apart that only one of them could still be true. The projector was showing the other one.

This is date reporting without dependency control: leadership still sees a fixed milestone and "progress" language while nobody owns a live predecessor graph — the tasks, durations, buffers, and handoffs that would force the date to move when the work underneath moves.

A schedule is a model, not a promise

The launch date is an output of the dependency model.

The model is dependencies, durations, and uncertainty.

When teams treat the date as fixed input and dependencies as optional narrative, recovery arrives too late.

Dependency-first schedule logic starts with:

1. required outcomes,
2. prerequisite tasks,
3. ownership of each task,
4. uncertainty ranges,
5. integration constraints between streams — where work in parallel has to meet (interface freezes, shared builds or fixtures, hardware–software bring-up order, qualification artifacts that block the next stream).

Then the planning lead computes the date from that model instead of declaring it in advance.

Honest critical path

Critical path is not “the work we care about most.”

It is the chain that controls completion date.

Rules for honest path management:

- recalculate path when task state changes,
- include external dependencies (supplier lead time — procurement and qualification clock from order or release to usable parts — plus compliance, tooling),
- include integration and validation tasks (not only build tasks),
- show float explicitly.

If your plan has no visible float and no uncertainty ranges, the schedule has no tested recovery path.

Failure modes that fake schedule confidence

Common schedule distortions:

- compressing non-critical tasks and calling it recovery,
- hiding blocked dependencies in "in progress" buckets,
- assuming parallelism where shared people/equipment make it serial,
- reporting roll-up percent complete while gate or sign-off prerequisites stay blocked — percent complete tracks effort and checklist breadth; decision readiness tracks whether the next honest gate, signature, or integration step is unblocked. Optimizing for the first lets teams paint progress green while prerequisites for the real decision are still missing.

These behaviors protect optics and destroy forecast quality.

◎

At an energy storage program, the announced launch date had not moved in three months on the program slide. Underneath, three thermal-related long-leads had quietly slipped: the cold-plate extrusion (11 working days, triggered by the thermal limit revision that changed the geometry requirement), the thermal interface material qualification (8 days, triggered by the contact-resistance finding), and a chamber booking for the revised configuration (6 days, because the test queue had moved). That geometry revision forced a new extrusion cross-section the original schedule had no entry for.

None of the slips were visible on the schedule because the schedule had been built date-to-date, not dependency-to-dependency.

When the team recomputed the chain — what depends on what, with honest durations and buffer for uncertainty — the new date was 19 working days beyond the announced date. That revision was discovered and communicated four months before launch, not two weeks before. The recovery conversation that followed was real: cut one product variant from the launch scope, change the supplier extrusion sequence to overlap with qualification, fund a parallel chamber booking. The conversation was hard. It was also the first honest program conversation in three months — four months of decisions that had been made downstream of a date everyone knew was wrong.

◎

Weekly cadence for schedule truth

A useful weekly cycle:

1. Refresh task state from owners.
2. Recompute dependency chain and current critical path.
3. Identify path deltas since last cycle.
4. Publish schedule delta (path and date impact) in the one-page status.
5. Trigger escalation where recovery needs scope/resource decision.

If the team does not recompute dependencies each cycle, the status packet is not decision-grade.

When the dependency chain is live and the weekly cadence publishes a delta note, the PM reads the path delta before the review instead of re-deriving it in the meeting. That is the whole reason to recompute every cycle: the schedule answers the status question before anyone has to ask it.

Acceptable homes for the dependency-based schedule are those where the program can rely on the same facts: one place everyone can open without hunting, a stable reference link from your program hub, predecessor links or an equivalent explicit dependency map that is maintained when work changes, and a visible history of plan churn (what moved, when, and why) so weekly deltas are not argued from memory. That can be a Gantt file, a table with a maintained dependency column, or a version-controlled outline — scale and ceremony should match program size. What cannot vary is that the date is derived from the dependency chain and uncertainty fields, not set first and defended by narrative compression.

What real recovery requires

Recovery has three levers:

- change scope,
- change sequence,
- change resources.

"Work harder" does not change dependency structure, so it cannot be a fourth lever.

For every recovery claim, require:

- which lever is being used,
- which dependency it changes,
- what risk it adds,
- who approved the trade.

Without these four fields, recovery is rhetoric.

Link to gates and one-page truth

Schedule integrity depends on records you already built:

- decision logs (what changed),
- risk register (what can slip next),
- one-page truth (what leadership sees),
- ownership map (who closes blockers).

If these are weak, schedule updates drift into negotiation instead of model updates.

If your plan has no visible float and no recovery lever named in writing, you do not have a schedule. You have a date with a story around it.

Field test: find the next major milestone on your program. Can you name the three real physical or procurement dependencies that drive that date, and point to their current values in a controlled record? If the date is a commitment rather than a dependency calculation, the schedule is a target, not a plan.

Part III - Technical evidence that changes program decisions

9. Requirements Lifecycle (Hypothesis -> Bless -> Evidence -> Revise)

Wednesday, 6:50 a.m. The thermal verification log shows the board thermal limit set to 65 °C, while the full-load chamber test at 40 °C ambient shows the control board measuring above 70 °C.

The program ran thermal management work — redesigning airflow, shifting components, iterating the layout — trying to bring the board temperature below the limit.

The requirement had a value and an owner. It did not have a scope note naming which operating scenario it applied to, or a rationale connecting the limit to actual component requirements.

Three weeks of committed design work against a requirement whose basis had never been confirmed.

Requirements are managed hypotheses

A requirement starts as a controlled hypothesis in the requirement log: this value must hold for a named operating condition. Both the value and the operating condition are part of the hypothesis. Changing the understanding of the bounding scenario — what worst-case actually looks like in field deployment — is itself a revision trigger, not only a measurement that crosses a threshold.

You make that hypothesis decision-ready by:

1. stating assumptions in the requirement record,
2. attaching bounded evidence to the same requirement ID,
3. recording controlled revisions with rationale,
4. assigning clear approval authority for each revision.

Treating early requirement values as permanent truth creates fake certainty and expensive rework once evidence arrives.

Lifecycle states (practical model)

Track each requirement ID through four lifecycle states:

- **Hypothesis**: initial value, assumptions explicit, low confidence.
- **Bless**: approved for current program phase with stated confidence/risk.
- **Evidence**: test/model/supplier data collected against requirement.
- **Revise**: requirement updated (or reaffirmed) with rationale and impact.

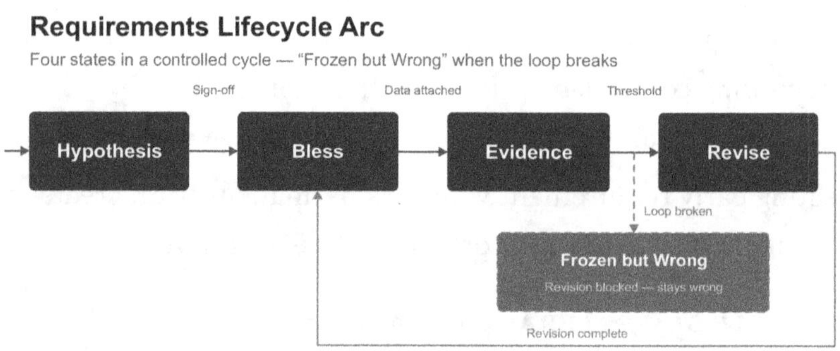

Figure 3: Requirements Lifecycle Arc — four states cycling through controlled evidence and revision; the dead-end "Frozen but Wrong" path shows what happens when the loop is broken

Each state requires an explicit move condition: Hypothesis advances to Bless on a named authority sign-off with stated confidence for the current phase. Bless advances to Evidence when the first data record is attached to the requirement ID. Evidence advances to Revise when data crosses a predeclared threshold or operating conditions change materially. Revise re-enters Bless once the revision record is complete — old value, new value, evidence source, rationale, approver, and date.

The requirement owner repeats this loop after each evidence update, gate review, or operating-condition change.

This loop makes changes legible in the log and mandatory for every downstream decision that consumes the value.

Frozen-but-wrong is worse than revised-with-evidence

Leads rightly fear uncontrolled churn in the requirement baseline. The opposite failure is also common: freezing wrong values in the requirement register to protect schedule optics.

A frozen-but-wrong requirement baseline causes:

- hidden divergence between design and test reality — the requirement log and the test record reference different values with no revision entry connecting them,
- late design turns — design is built to a known-wrong bound until a gate exception forces the change after commitment,
- supplier misalignment — supplier inspection criteria reference the frozen value, not the test-confirmed limit,
- and "surprise" gate failures — visible earlier in data, masked by a requirement record that stayed frozen.

The lead needs both controls in the requirement log: revision discipline stops uncontrolled churn, and evidence discipline stops the opposite failure — holding a known-wrong value to protect the appearance of a stable baseline.

◎

The board thermal limit was set to 65 °C by the electrical engineering team — a conservative estimate intended to protect component lifetime across the product's operating range. It was not decomposed to the case temperature limits of the specific components on the board, and no scope note identified which operating scenario it applied to.

The full-load chamber test ran at full power and 40 °C ambient — the scenario defined for UL qualification — and measured the control board approaching 70 °C. The limit said 65 °C. The program ran three weeks of thermal management work: airflow changes, layout iterations, component placement. None of it was targeted, because the requirement had not been tied to the actual components at risk.

When the requirement owner pulled the 65 °C limit back to its basis, two questions surfaced: which components it was protecting, and which scenario it applied to. The answers changed the problem. The 65 °C estimate was a lifetime operating condition limit — it applied to sustained load in worst-case ambient across the product service life, not to a one-time UL qualification test at 40 °C. No component case temperature limit was being violated under the qualification test scenario. For the lifetime scenario, the question became which specific components were at risk and what their case temperature limits were — a decomposed set of requirements the 65 °C proxy had been standing in for.

Three weeks of whole-assembly thermal management work was redirected to targeted mitigation on the two components whose case temperatures were actually at risk under lifetime operating conditions. The target list had two components on it. The previous three weeks had addressed neither.

The fix was not a thermal redesign. It was the sentence that should have ridden alongside the number the day it was written: which scenario, which components. Two clauses, three weeks late.

Revision criteria (minimum)

Revise a requirement only when predeclared decision thresholds are met:

1. evidence crosses a pre-defined threshold,
2. operating conditions changed materially,
3. supplier/process capability data invalidates prior assumption,
4. a pending gate decision or design release depends on confirming or correcting the current value before commitment.

Each revision entry in the requirement log should include:

- old value, new value,
- evidence source,
- rationale,
- impact scope (design, test, cost, schedule),
- approver and date.

Acceptable implementations for the requirement lifecycle log include a gated product-data revision route when build truth must trace to an authoritative BOM, an issue-tracker type with gated fields for owner, evidence links, and revision trail when releases are routed from backlog records, a locked shared spreadsheet with named owners and protected history columns for early programs that still need fast iteration, or a controlled markdown log under version control when the team is small and repository-led. Whatever the home, the same row has to carry all of it: old value, new value, evidence source, rationale, impact scope, approver, and date.

If the revision row exists but the evidence source field is blank, the revision is not controlled — it is a number change with a timestamp. That is the tell.

Preventing moving-goalpost behavior

Teams resist revisions when the decision record looks arbitrary.

Prevent goalpost behavior by separating two written records:

- **decision criteria** (requirement ID, revision threshold, and decision owner — defined and locked in the revision memo before data arrives),
- **decision outcome** (revised value or reaffirmation — recorded after criteria are evaluated, with evidence source and approver).

If criteria stay stable in the decision record, revisions read as disciplined learning rather than political drift.

Weekly lifecycle review

For each active high-impact requirement ID, run a short weekly review:

- state check (Hypothesis/Bless/Evidence/Revise),
- evidence delta since last review,
- pending revision decisions and owners,
- downstream artifact update status.

The requirement owner should finish this review in minutes and publish one delta note, not run a half-day meeting.

Field test: If you cannot point at one requirement ID in your current program that revised in the last quarter with evidence in the log, an approver on record, and a downstream-update note, you are running on frozen-but-wrong values somewhere. Re-read *Frozen-but-wrong is worse than revised-with-evidence.*

10. Requirements from Physics (Not Stacked Buffers)

By mid-morning Wednesday the structural load requirement carried three separate safety factors — one from the component lead, one from systems, one from operations. The rationale table had a row for the number and no row for what uncertainty each factor was covering.

No owner could say which was which. The factors had stacked over time, each defensible on its own, none traceable to a named risk.

The released design came out 6% heavier, material cost up, pilot-line yield down against the previous baseline.

That is unmanaged uncertainty in the requirement line, presented as conservatism with no one accountable for any single part of it.

◎

Physics-first requirement logic

Start from a measured physical limit in the requirement record, then add an explicit program buffer with an owner.

Write each requirement as a three-layer decision object:

1. **Define physical limit** (what nature allows under named conditions),
2. **Set engineering target** (what design commits to hit),
3. **Declare program buffer** (what uncertainty margin is carried, by whom, and why).

If those layers are merged into one opaque number, no decision owner can tune risk intelligently at gate review.

How stacked buffers happen

Buffer stacking usually starts with three good-intention moves:

- component lead adds margin in the requirement for model uncertainty,
- system lead adds margin for integration uncertainty in a downstream spec,
- operations lead adds margin for process uncertainty in release criteria.

Each margin looks reasonable in isolation, but combined they can exceed the product's weight, cost, or manufacturability limits.

◎

At an energy storage program, three teams each added a thermal margin to the switching-stage requirement. The thermal lead added 5 °C for component tolerance. The systems lead added 3 °C for ambient worst-case. The manufacturing lead added 2 °C for process variation. All three margins were honest engineering judgments. Combined, the requirement was 10 °C tighter than any single physical driver justified — and the chosen architecture could not satisfy it. When the teams traced each margin to its physical source and its uncertainty basis, 7 °C of the buffer was recoverable without increasing real risk. The requirement moved from unmeetable to achievable without changing the architecture — and a two-month design study for a thermal architecture change became unnecessary once the margins were untangled. The conversation that untangled the three margins took ninety minutes. Eight weeks of design study replaced by a traceability column.

◎

Fake safety vs real safety

More margin in a requirement line is not automatically safer for the delivered product.

Unexamined margin in the requirement can:

- push design into new failure modes (mass, thermal, packaging) during integration,
- force expensive materials or processes in sourcing,
- reduce manufacturability and yield on the pilot line,
- hide which uncertainty actually needs test closure before gate sign-off.

Real safety is explicit risk control — named owner, named uncertainty, named evidence. What you have when you stack unexamined margin is a tighter number and no idea how much of it you actually need.

Buffer hygiene rule

Every requirement buffer must carry five explicit fields:

- assign a named owner,
- identify the uncertainty source it covers,
- state a quantitative value,
- define a specific expiration or review trigger (a named gate, a test result, or a date — not "as needed"),
- link planned evidence that will reduce or confirm it.

If a buffer has no owner or retirement trigger, it persists by inertia and silently hardens into baseline policy.

You have watched this happen. The 2 °C process buffer from the first design phase is still in the requirement four builds later, and nobody can find the engineer who added it.

Quick variation coverage check

Before blessing a requirement ID, run this variation-coverage check:

1. Which variables drive this limit most?
2. What ranges are assumed for each variable?
3. Which interactions matter?
4. What evidence currently supports those ranges?
5. Which assumptions are weakest and need targeted test/model work?

This review does not require advanced math in the meeting room; it requires explicit assumptions in the requirement record.

WRITING A CLEAN REQUIREMENT LINE

Weak requirement line: "Part shall withstand expected thermal load with margin."

Better requirement line: "At condition set X/Y/Z, measured by method M, value shall be <= N. Program buffer B covers uncertainty U, is owned by O, and is reviewed at gate G."

The intent is the same, but the owner, uncertainty source, and gate review trigger are now explicit and auditable.

Field test: If you can name the physical source and the uncertainty basis for each buffer in your tightest requirement, the requirement is grounded. If you cannot, the margin is ungrounded — stacked buffers are the most common reason, but the consequence is the same regardless of cause: you do not know how much design margin you actually have.

11. Many Factors, One Honest Limit

The subsystem passed every single-factor check in the verification report. On paper it was green.

Then the combined-condition run — every variable loaded at once, the way the real product sees them — failed the requirement boundary. The team had tested one factor at a time and never run the combined envelope.

By Wednesday afternoon the gate owner had a stack of passing single-factor data and no basis to choose. The test plan had burned schedule without lowering the uncertainty that actually mattered.

When one-factor reasoning breaks

One-factor testing is useful during early screening. It fails when interaction effects dominate the decision boundary.

Use these trigger signs to decide when one-factor logic has broken:

- integration repeatedly reveals failures not predicted by subsystem tests,
- build-to-build results diverge without a single dominant cause,
- response is highly sensitive near operating boundaries,
- multiple coupled assumptions sit inside one requirement decision.

EMI is a common example: you can model the source with good accuracy, but coupling to other structures depends on layout, ground return paths, and shield geometry all at once. Shielding effectiveness calculated for a single source is only useful if you have correctly identified every significant source. Early testing with physical hardware can locate the coupling paths that single-domain analysis missed; catching that late means architectural changes after routing and mechanical design are fixed.

If any of these signs are present, additional one-factor testing usually burns time without reducing uncertainty enough to support a real decision.

Lead's job: scope the decision question

The lead does not need to run the analysis personally. The lead must scope a decision-ready ask with a named owner and due date.

Use this minimum ask format for the analysis request:

1. Name the decision to support,
2. cite the impacted requirement,
3. define factors and ranges to study,
4. specify output needed (limit, window, or pass/fail region),
5. set confidence/evidence threshold,

6. assign date and owner.

Without this framing, teams can generate impressive activity artifacts that still produce no bounded conclusion for the decision log.

◎

A heat-generating module was mounted to a cooling structure. The thermal model predicted it would stay within its limit at full operating load. Three early prototype builds ran consistently hotter than predicted – not scattered results, the same gap every time.

The root cause was in what neither model was told to talk to the other about. At room temperature, the mounting assembly held good contact pressure across the thermal interface. At operating temperature, differential expansion between the module and the frame reduced that pressure at the interface edges. Lower pressure meant less effective contact. Less effective contact meant higher thermal resistance than the thermal model had assumed as a fixed input. Higher resistance meant higher temperature – which drove more expansion – a feedback loop no single-domain model could see.

Each model was valid given its own boundary conditions. The thermal model used a room-temperature interface measurement. The mechanical model validated contact at room temperature only. Neither was wrong. The interaction between them was the missing factor.

Fix: simplified geometry prototypes with one variable controlled at a time, each modeled to match before adding the next. The coupled model, validated against those steps, reduced prediction error from a consistent overshoot of roughly one temperature class to within a few degrees at full load. The same discovery post-tooling would have required a geometry change to the mounting architecture.

Rejecting activity theater

Reject analysis summaries that report:

- "we tested a lot" without decision mapping,
- "model looks good overall" without requirement impact,
- "results are mixed, more work needed" without closure criteria,

because none of those statements names a decision owner, requirement artifact, or closure trigger.

Accept analysis packages that answer:

- what changed in the decision baseline (the program's current standing commitments),
- what uncertainty remains and who owns it,
- what risk moved in the register,
- what next decision date is now justified.

What "honest limit" means

An honest limit is not the most conservative number anyone can defend. It is the best current boundary supported by evidence, explicit assumptions, and named risk ownership.

Document the honest-limit package with four fields:

- define tested or modeled domain,
- state interpolation or extrapolation assumptions,
- report confidence bounds,

- list known blind spots.

This keeps confidence proportional to evidence when the gate chair reviews the limit decision.

Tie outputs to program controls

Analysis results are only useful if they propagate into program artifacts:

- update the requirement value or buffer basis,
- update the risk register entry,
- update the one-page status delta,
- feed the next gate decision packet.

If outputs stay in a technical slide deck, no program control artifact changes and behavior stays the same.

Thin bench approach

If the team lacks depth in multi-factor analysis:

- borrow specialist support for method execution,
- keep the decision question narrow and dated,
- require output in program language, not only specialist language.

Lead responsibility remains: the lead still owns the decision question, artifact quality bar, and closure date.

Field test: name the analysis your program is waiting on right now. Does the request include the decision it is meant to support, the factors and ranges to study, and the date the decision is needed? If not, the analyst is guessing what matters — and the result will not close the decision, only inform it.

12. Models and Tests That Change Decisions

Wednesday, 6:40 p.m. The decision review packet holds 20 model runs and 12 test reports and still cannot resolve the thermal case temperature requirement.

The gate decision still hangs because no owner can show baseline, delta, and effect size in one traceable artifact.

Evidence volume is high, but decision power stays low when each run is not linked to a baseline requirement state. The missing discipline is not more runs — it is calibrating the model against the chamber and expressing the result as a delta from baseline: what changed, by how much, and what that means for the open requirement. That calibration-and-delta loop is what this chapter installs.

Evidence must be delta-first

Your model or test result is useful only relative to a named baseline artifact — without the baseline, you cannot tell whether the new run changed anything or just confirmed that you ran it again.

For every run record, capture a decision delta package:

1. cite baseline reference,
2. state what changed,
3. predict expected effect,
4. report observed effect,
5. declare decision implication.

Without this package, results become isolated artifacts and cannot form a learning sequence that changes decisions.

Naming and traceability discipline

Use one boring but enforceable traceability convention:

- keep stable requirement ID linkage,
- assign run and test IDs,
- record revision IDs for design and setup,
- stamp timestamp and owner,
- store records in a location with immutable history.

This discipline prevents decision debt: review leads pay the cost at gate when they cannot compare outcomes across builds and teams.

Pair model claims with physical evidence

Models are strongest when analysts calibrate them against physical evidence. Tests are strongest when reviewers interpret them with model context and assumptions.

Use these pairing questions before changing a requirement value:

- Where do model and test agree?
- Where do they diverge?
- Which assumptions explain divergence?
- What decision can be made now despite remaining uncertainty?

The goal is not perfect correlation before every move; the goal is honest confidence for the next dated decision.

A thermal model for a power-electronics module predicted the switching-stage case temperature would hold at 73 °C at full duty — safely below the 75 °C thermal case temperature limit. The chamber test measured 78 °C steady-state. The team's first interpretation was that the chamber was measuring something different than the model. The calibration finding was narrower: the model's contact-resistance boundary condition assumed a thermal-grease application that the actual mounting jig could not deliver. The jig had been designed for a different module with different interface geometry — it had been on the test floor for three years, and nobody had updated the boundary-condition assumption when the module changed. The model was correct given its boundary condition. The boundary condition was wrong given the actual assembly. Treating the model and the chamber as independent confidence votes is what stretched this out: a full week of "the chamber must be wrong" before anyone examined the boundary condition. One boundary-condition correction and a re-run later, the model predicted 79 °C — matching the chamber within measurement uncertainty, a delta of one degree where there had been five. What changed was not the model and not the chamber but one entry in the boundary-condition log: the contact-resistance value, now traced to a direct measurement on this module instead of carried over from a different one. The requirement revised to 78 °C — a controlled update: the correction is the evidence, the new value is documented with a named owner, and the baseline moves.

That revision is a controlled evidence update — the requirement baseline moves only when evidence satisfies predeclared criteria and the rationale is documented.

◎

Evidence threshold for requirement changes

Do not revise a requirement on a single model run that shows a passing result.

Set requirement-change thresholds in advance:

- confirm the test or model ran at the operating conditions the requirement specifies,
- verify repeatability is acceptable,
- show sensitivity to key variables is understood,
- bound measurement uncertainty enough for the decision at hand.

If threshold is not met, the evidence owner documents what is missing in the review artifact and sets a closure date.

Common evidence failure patterns

- orphan files with no requirement linkage,
- results impossible to reproduce from metadata,
- "best run" selection without rationale,
- model updates with no calibration note,
- test deltas reported without setup changes.

If you can tick two or more of these in your current evidence folder, the gate review will be a debate, not a decision.

Each pattern inflates confidence in the artifact while leaving decision uncertainty unresolved at review.

Practical evidence package for decisions

For any decision review, provide one compact decision package:

- state question to answer,
- summarize baseline and delta,
- rate evidence quality,
- recommend a decision,
- assign residual risk and next evidence step.

This package keeps technical depth available while giving exec and PM reviewers a clear, fair decision frame.

The boundary-condition correction is confirmed. The model re-run matches the chamber within measurement uncertainty. The gate decision changes from hold to proceed — one week after the chamber result that everyone initially attributed to measurement error. The requirement is 78 °C, the model predicts 79 °C, and the boundary-condition log now carries the contact-resistance value traced to the direct contact-resistance measurement. The decision packet is one page. The technical backup is three.

Field test: name one model your program relies on. Did the last model update change a decision, or did it confirm a decision already made? If it confirmed what was already decided, the model is not driving decisions — it is defending them. Find the last result that was acted on and trace it to the decision log.

13. Fleet as Instrumented Experiment

Thursday morning: five units on the bench, and no two run logs structured the same way.

By Friday the team had a story about each unit and no way to compare them – which build carried which change, which result came from which configuration, none of it lined up.

That is a missed experiment, not a successful fleet. The distinction matters because the fleet's job is not to confirm a prototype – it is to surface variation across units so the risk register moves on evidence, not on which unit the program lead happened to visit last. Getting there requires structured per-unit records, a build matrix that controls variables, and a weekly cadence that forces comparison. Build those three and the fleet becomes an experiment; skip them and it stays a pile of units with good stories attached.

Fleet mindset: learning asset, not milestone artifact

A fleet build is valuable when each unit produces usable evidence.

Treating prototypes as proof-by-existence wastes the most expensive learning window in the program.

Fleet intent should be explicit before build:

1. Which decisions this fleet is meant to inform,
2. Which requirements/risks each unit is probing,
3. Which measurements and conditions must be captured.

A unit's evidence is gate-complete when the record answers all three from the record alone — without asking the engineer who ran the test.

Minimum per-unit record

For each serial unit, capture:

- configuration (revision/state),
- manufacturing context (process/material lots),
- test conditions,
- results against requirement IDs,
- anomalies and disposition,
- owner/date for record completeness.

If there is no per-unit record, you cannot defend the cross-unit comparison at gate — and the anomaly that matters will be in the unit whose log is thinnest.

Six fields. If the record cannot survive the question "what was different about this unit," it is not a unit record — it is a bench note.

Build matrix, not pile of units

Plan fleet variation deliberately:

- what to hold constant,
- what to vary,
- which interactions are intentionally sampled,
- which units are designated for destructive/edge tests.

If all units are "nominal," fleet learning is shallow and late surprises remain likely. A common miss: a five-unit fleet built to identical nominal conditions produces five confirmations of one operating point rather than a sample of the variation the program will face at scale.

Connect fleet data to program controls

Fleet evidence earns its keep only when it moves the same program-control records every result should — requirement confidence, the risk register, the one-page status, and the next decision in the queue. When the chain runs, that propagation is visible: Unit 03 measured a leak rate well above Unit 01 at hot soak, the seal leak requirement confidence moved to red, and the next gate call shifted from go to hold pending seal redesign evidence.

If fleet data stays in test notes, the program keeps running on stale assumptions.

◎

At a battery-adjacent program, five prototype units were built and instrumented over four weeks. Unit 03 measured a leak rate more than twice Unit 01's baseline at hot soak — not a sensor artifact, the same delta confirmed across multiple test runs. The discrepancy sat in the test engineer's notes for eleven days before it reached the program's risk register, because there was no per-unit record requirement that forced the delta into the risk log. The test engineer had mentioned it verbally on day three. It was in nobody's written record. The thermal requirement moved to red only after the team had already placed tooling orders on the assumption that Unit 01 was representative.

The program added a per-unit record requirement and a weekly fleet review cadence to its standing controls. The next fleet campaign surfaced three requirement deltas in the first review week rather than the fifth. That gap had cost tooling decisions made on unrepresentative data — decisions that required a correction order once the real distribution of results was visible.

◎

Failure patterns in fleet execution

Common misses and what catches each:

- unit history lost after rework — per-unit record requires a rework entry before the unit re-enters evidence
- instrumentation setup drift between units — per-unit test-conditions field, visible at the evidence completeness check
- anomalies documented but not tied to decision owners — step 3 of the weekly review (unresolved anomalies and owners)
- test logs decoupled from build configuration — step 1 of the weekly review (unit-by-unit evidence completeness check)

Without the pairing, these failures make apparently rich data unusable for high-stakes decisions.

Practical weekly fleet loop

During fleet campaigns, run a weekly review:

1. Unit-by-unit evidence completeness check,
2. requirement/risk deltas from new data,
3. unresolved anomalies and owners,
4. decision queue updates for next gate.

Short, strict, repeatable. Each cycle ends with a documented delta note: which requirement or risk rows moved and what evidence moved them.

Delta discipline scales to fleet

Delta and traceability discipline at component and subsystem level scales directly into fleet work — the same logic, applied across physical units in real variation.

When done well, fleet turns uncertainty into bounded decisions before major spend.

Field test: If your weekly fleet review this week can name one risk-register row that moved because of a specific unit's measured delta, the campaign is doing its job. If it cannot, you are collecting prototypes, not evidence.

14. Risk and How We Define Done on Paper

The gate review package marked every action closed.

Two critical failure modes were still open in the lab log — verification and reliability tracking both, with no accepted closure evidence. It was Thursday, gate prep underway.

The paper package says done, but the program state is not. The gap is not a communication failure — it is a structural one. The action tracker and the failure-mode log are not the same artifact, and no gate-packet requirement connected closure of one to the other. Fixing it requires a single rule: closure evidence must cite the failure-mode investigation disposition, not just the action ID. That single rule is the spine of this chapter.

That gap is where late cost, schedule slips, and credibility damage enter through integration and executive review.

This is the gate-closure logic that depends on calibrated chamber evidence and a controlled requirement revision — in the canonical case, the closure that must happen before the thermal case temperature requirement can close at the release gate.

"Done on paper" must match decision reality

Artifacts are useful only when a named owner uses them to change a specific decision and downstream behavior.

A credible "done" statement is a decision object that must:

1. state explicit criteria (gate chair's pass condition),
2. cite evidence references (gate chair's confidence source),
3. include owner sign-off,
4. declare residual risk (gate chair's accept-or-escalate input),
5. confirm downstream updates are complete.

If any element is missing, "done" is only a status label and fails as a decision condition at gate.

Minimum credible artifact standard

For each risk or verification artifact, require this minimum metadata:

- name artifact owner,
- record revision and date,
- link requirement or risk ID,
- state decision linkage (what choice this supports),
- report current confidence and known gaps.

If the artifact has no linkage, the gate chair cannot use it for a decision.

Artifact evolution across builds

Risk artifacts should evolve with new evidence across builds, not reset at each phase boundary.

Each artifact revision should show four deltas:

- state what changed,
- record what remains uncertain,
- retire concerns with evidence and rationale,
- declare what new risk emerged.

Resetting artifacts to "clean" each phase destroys institutional memory that gate and program leaders need.

Busywork vs decision work

Reject artifact packages that:

- are completed only to satisfy format,
- contain generic failure text with no operating context,
- omit a named owner or closure date,
- cannot point to a decision they changed.

Keep artifact packages that:

- expose uncertainty honestly,
- map to active decisions,
- summarize tradeoffs in language each tier can evaluate without follow-up questions.

You know the difference after the first gate debrief where someone produced a risk summary that nobody had read. That is not a documentation gap — that is what busywork looks like when it reaches the table.

A gate package at a large industrial program marked all thermal actions closed. Two critical failure modes — thermal runaway propagation and connector derating under field humidity — were still under active investigation in the test lab. The disconnect: the action tracker was PM-owned and tracked action-item status by ID. The failure-mode log was engineering-owned and tracked investigation state by failure mode. No artifact required a closure entry to cite a failure-mode log entry rather than just an action ID. One internal audit caught the gap three days before the gate review. The auditor had not been looking for it specifically — she was tracing an action item listed as closed with no linked evidence, and the failure-mode log was the first place she checked. If the gate had proceeded, the program would have released a build configuration while two unresolved failure modes were still in active investigation — a six-to-eight-week integration cost if either materialized in production. The fix was one line in the gate-packet template: closure evidence must cite the failure-mode log entry and its investigation disposition, not just the action ID. The organization had been closing gates the old way for two cycles and had shipped at least twice with failure modes still open. Three days of one auditor's scrutiny was all that separated this gate from the same outcome.

Tie-ins to Gates and One-Page Truth

You need risk artifacts to feed the records that gate decisions depend on. When they stay in a technical silo, leadership walks into the gate review with paper confidence that nobody in the lab would recognize.

They must feed the decision records the gate depends on — the gate decision itself, the risk register, the one-page narrative, and escalation to named owners when a failure mode stays open. If this propagation is weak, paper confidence drifts from the real program state and misleads leadership decisions.

What matters is that one record exists, everyone knows it is where failure modes and their investigation status live, and it is the record that gets pulled for the one-page summary and the gate package — not a status assembled fresh for the occasion.

Practical "done" review in 15 minutes

For each high-impact closure item, ask:

1. What was the done criterion?
2. What evidence satisfies it?
3. What risk remains despite closure?
4. Which dependent decisions changed because of this closure?

If questions 3 and 4 are blank, closure is cosmetic — do not treat as done; return the item to the evidence owner before the gate proceeds.

Done means the failure mode is dispositioned, not the action item is closed.

Field test: find one item your last gate called closed. Can you name the evidence that moved it to closure, the DRI who accepted residual risk, and the date that decision was recorded? If the closure record has none of those, it is closed on paper, not closed in fact.

Part IV - Running the Organization

15. What Every Tier Wants (Exec to Supplier)

Thursday, 12:55 p.m., lunch cleared to the corner of the table, the weekly review still going.

The exec pulled the status from last week's board deck. The PM has the project tracker open on one screen. Engineering is working from an email thread that captured the last three decisions. The supplier's rep is at the table with a spec revision they received three weeks ago, which has since been superseded twice.

Nobody is wrong about what they brought. Each person is reading a legitimate artifact — a written program record. The problem is that the artifacts are not the same artifact.

The ninety-minute review stalls forty minutes in because a decision that everyone assumed was made — a packaging change affecting the supplier — had never been recorded in a form any two tiers could read together. The exec asks for a confidence number. Engineering needs two more weeks of closure time. PM needs a stable input to forecast. The supplier needs one released package with one owner. All four needs are legitimate. None of them can be met in this meeting because the four tiers are navigating by different maps — on average, a decision and a half per week carried this way, each adding two to five days of downstream delay.

Alignment fails at interface contracts

Most cross-tier conflict is not ideology. It is interface-contract ambiguity: what each tier needs to make a decision, when that input is needed, and which quality bar applies to the artifact. When those three things are not written down, every tier improvises. Improvisation looks like friction.

When a named owner publishes these contracts in one shared artifact, the friction does not disappear — but it becomes auditable. An auditable gap gets assigned and closed. An invisible gap becomes a personality conflict.

What each tier usually needs

Keep each tier contract plain, operational, and tied to one decision artifact.

Executive / sponsor - leading indicators of risk and dependency — not a confidence score, but a named risk row with an owner and a trigger - clear decision asks — one binary question with a date, not a discussion topic - tradeoff framing when date/cost/scope conflict — their job is to make that call, not to infer it from color-coded slides

What the exec owes downward: stable goals that do not shift weekly, in-room tradeoff decisions when engineering brings the evidence (not hallway overrides), day-grade resource unblocking when the engineering lead names a blocker, and a clear definition of "done enough to ship." An executive who takes the decision asks and the tradeoff evidence without closing the call keeps the risk item open. It will run through every weekly review unchanged — re-reported, re-acknowledged, unresolved — while the work that needs that decision queues behind it.

PM - stable owner map — who has each decision, one row per open item - decision-ready updates — the PM should not have to chase an answer; the owner posts it before the review - early signal when assumptions break — if the PM hears about a schedule shift from the exec, the information routing is broken

What the PM owes downward: transparent escalation when a contract breaks (not a complaint — a named artifact gap, a blocked decision, a date), dependency hygiene in the one-page update, and a cadence that does not add overhead faster than it removes ambiguity.

Engineering lead - clear requirement and risk state — one live record, not email - authority boundaries — which decisions belong to the lead and which need approval; ambiguity here is expensive - protected closure windows for technical decisions — time to think is not a luxury; a lead with no protected windows will make shallow calls on deep problems, because the problem context resets at every interruption

Manufacturing/operations - unambiguous release package with revision-locked content - process-critical characteristics called out explicitly, not scattered through engineering notes - change notice with timing — not just "we changed it" but "this revision affects your build starting with lot N"

Supplier - one current revision source — not "check the portal, the email, and the drop folder" - expected evidence and acceptance criteria before the program needs the data, not when inspection fails - single owner for both technical and commercial closure – when these run as separate conversations, the supplier receives two partial agreements and acts on whichever arrived last; no one on the program side holds the combined answer, and the gap surfaces in the release package

Reciprocal obligations

Needs without reciprocal obligations become entitlement and eventually gridlock. The contract is not "here is what I need from you." It is "here is what I need, here is what I deliver in return, and here is the artifact that makes both visible."

Pair each ask with an owed output artifact:

- Exec asks for confidence → owes timely trade decisions recorded in the decision log and stable goals that do not drift week-to-week.
- PM asks for predictability → owes transparent escalation and dependency hygiene in the one-page update, and a cadence the team can trust.
- Engineering asks for room to solve → owes explicit risk state and closure dates in requirement and risk records, and a commitment to surface surprises before the weekly review, not at it.
- Supplier asks for clarity → owes capability evidence and issue response cadence in the release package.

This framing keeps fairness visible by balancing each tier's ask with a concrete, auditable obligation.

PM fairness, explicitly

When record quality is low, PM becomes the human sync engine compensating for missing system behavior in formal artifacts. The

PM pings because the artifact does not answer the question. That is a system failure, not a PM failure.

With clean ownership, requirement truth, and one-page integrity, PM work shifts from chasing contradictions to clearing real blockers. That shift has a concrete payoff for the engineering lead: **a PM who trusts the schedule and the risk register stops asking for daily status.** Not because they stopped caring, but because they have something reliable to read. An engineering lead who has earned that trust has bought themselves a working environment — time to solve the technical problem instead of presenting it repeatedly.

That is the contract to protect. An efficient PM, backed by a process that does not waste their time, is a PM teams respect — and that earned trust is itself the measurement that the artifacts are doing their job.

What the exec leaving early looks like

There is a simple field test for whether the program's one-page status and gate truth are working: **can your exec leave for the weekend without anxiety?**

Not because the program has no risk. Every hardware program has risk. But because:

- the risk register surfaces the risks that would otherwise surprise them
- every open decision has a named owner and a date
- the one-page status line traces to a real artifact, not to a slide that was polished to look calm
- the gate status reflects what the test floor actually saw, not what the team hoped to see

When those four things are true, an exec who leaves Friday afternoon is not checking out. They are making a rational decision that the information they need will surface through the artifact, not through a weekend phone call. That is the payoff for having built the OS. The fact that the exec can leave early is the measurement.

If your exec calls you on Saturday, the OS has a gap. Find it.

One review format that surfaces gaps early

Run a short cross-tier interface review against one shared contract sheet — one page listing what each tier needs and delivers each week, with a named owner per row. Not a status update: a contract-quality check:

1. Confirm what each tier needs this week (read from the contract sheet, not from memory).
2. Verify what each tier received (not what was sent — what actually landed in usable form).
3. Identify where contract quality failed (which artifact was missing, late, or incomplete).
4. Assign which owner closes the gap by when.

Skip long debate. Focus on interface quality, owner assignment, and dated closure commitments. The meeting produces a short gap record: gaps found, gaps owned, gaps closed. That record is the input to next week's check.

Escalation becomes cleaner

When tier contracts are explicit, escalation becomes factual and auditable: identify which contract failed, state what decision is blocked, quantify the current risk and date impact, assign who must decide by when.

That path resolves blockers faster than personality-driven escalation because decision authority and artifact gaps are explicit. An engineering lead who escalates with a named artifact gap and a dated decision need is not complaining — they are running the process. An exec who receives that escalation has a job to do, not a people problem to manage.

Field test: name the most recent decision your exec made on your program. Did they have a one-page status that traced to a live record? If they had to call someone first, the artifact was not trusted — and the call is now the record.

16. First Article: Fast, Disciplined Iteration

The first-article build ran three days before anyone clocked that the problem was not the hardware. It started rough, as expected; every shift made "small improvements." By Thursday afternoon yield was still bouncing and no one could say which change had moved it.

Three days went into root-cause investigation that found nothing wrong with any single change — only that the configuration was never locked between shifts, so every build started from a different unknown baseline. Throughput had climbed the whole time. The run log had quietly stopped producing anything you could decide from.

The team locked the build configuration at shift handoff: one baseline, one change per build cycle, signed off at handover. The first run log produced under that discipline answered two questions the previous log could not — which change moved yield and which did not. Root-cause investigation eliminated. The throughput was lower than the unlocked approach for the first two cycles; the learning was usable starting from cycle one.

Objective of first article

First article is not a demo milestone.

It is a structured learning phase that reduces uncertainty fast enough to decide next commitments.

Primary objectives:

1. validate critical functions under real build conditions,
2. expose process and design failure modes early,
3. convert findings into controlled changes with measurable outcomes.

A run log that carries decision-ready findings before the next commitment is made is the observable evidence that the phase ran as learning, not throughput.

If the phase optimizes only for throughput, it hides the very information you need most.

Iteration loop that actually learns

Use a repeatable cycle:

1. Define current failure/constraint.
2. Choose one controlled change set.
3. Record baseline and expected effect.
4. Run and capture results.
5. Decide: keep, revert, or escalate.

Then repeat on the next highest-risk constraint.

This loop keeps iteration speed high while preserving a causal record for decisions.

CHANGE DISCIPLINE RULES

Default rule: one-variable change where practical.

When multi-variable changes are necessary, require:

- explicit rationale,
- expected interaction risk,
- rollback condition,
- ownership of interpretation.

Without this, first article becomes untraceable trial-and-error.

LOGGING STANDARD FOR FIRST ARTICLE

For each iteration, log:

- unit/build ID,
- baseline state,
- change introduced,
- result against criteria,
- anomaly notes,
- decision and owner.

If logs are optional, memory becomes the system and confidence outruns evidence.

COMMON ANTI-PATTERNS

- unlogged changes between runs (change-introduced field blank or absent),
- lost baselines (baseline-state field blank at run start),
- vague pass/fail definitions (result-against-criteria field contains narrative instead of a named criterion),
- "good enough" closure (decision field names no predeclared threshold, only a judgment),

- change accepted while anomaly notes remain open (decision field reads "keep" while anomaly-notes field is unresolved).

These patterns feel energetic and usually lengthen ramp.

Decision gates inside first article

Do not wait for formal program gates.

Use mini-gates within first article:

- Is the change repeatable?
- Did risk move enough to proceed (relative to the predeclared entry criterion or the prior-run baseline)?
- Do we need deeper analysis before next run?
- Should scope or sequence change now?

This protects speed while preserving decision quality.

Thursday, second week, same time of day. The change log is posted on the workbench before the next shift arrives. The incoming lead reads it in five minutes, starts from a known baseline, and when yield drops on run 14, the delta investigation takes twenty minutes — not three days — because the configuration is locked and the prior decision is on paper.

Field test: find your last first-article result. Is there a written learning extraction that links specific findings to specific design or process changes? If the findings fed the conversation but not a written record, the iteration is fast but not disciplined — and the next yield drop starts from zero again.

17. Supplier Data, Critical Characteristics, and Producibility

"The supplier says the process is stable, so we're good." Friday morning, the supplier package spread next to our own incoming Cpk charts, and that sentence was the whole basis for the schedule.

The incoming variation said otherwise. Nobody could tell whether the gap was process drift, measurement mismatch, or requirement ambiguity — and "the supplier says it's fine" is not one of those answers.

Until someone owns that discrepancy and closes it, the schedule claim is unsupported.

Supplier truth must be operationally comparable

"Supplier says it is fine" is not evidence.

You need comparable evidence across teams:

- same characteristic definition,

- same measurement method or cross-calibrated method,
- same acceptance criteria,
- same revision baseline.

Without comparability, disagreements are endless and closure is slow.

Critical characteristics: keep the list tight

When teams label too many characteristics as critical, control plans lose focus and detection degrades.

Define a focused critical-characteristics set based on:

1. safety/regulatory impact,
2. function/yield impact,
3. downstream rework cost if missed.

For each characteristic, specify:

- nominal/limits,
- measurement method,
- sampling plan,
- owner on both sides (internal and supplier),
- revision baseline (aligned with the current controlled spec revision).

A tight characteristics list answers what matters; the evidence package determines whether the supplier can deliver it before the program commits.

Minimum supplier evidence package

Before major commit, require:

- latest controlled spec alignment confirmation,
- capability evidence on critical characteristics,

- measurement system notes,
- recent drift/anomaly history,
- containment and corrective-action status for open issues.

Require only the evidence package needed to make and record the commit decision.

Capability discussions in plain language

Do not let reviews drift into slide-heavy statistics discussion that delays a release decision.

Ask practical questions:

- Can this process repeatedly hit the required window?
- Under what conditions does it fail?
- How quickly do we detect and contain drift?
- What is the agreed response when it drifts?

Keep derivations in backup; keep pass/fail decisions in front.

Escalation playbook for discrepancies

When supplier/internal data conflict:

1. freeze to known-safe operating state if needed,
2. align revision and method baselines,
3. run short joint verification plan with owners/dates,
4. decide containment vs release based on agreed criteria,
5. update risk register and one-page status.

Delay usually comes from unclear ownership at step 3. The playbook's other function is framing: a discrepancy resolved through the five steps is a process gap with an owner — not a supplier failure with a consequence.

A program sourcing custom precision components got the same answer from every supplier when asking about process tolerance: the process held a stated tolerance regardless of feature size. Multiple suppliers, independently, the same number. It had the feel of an industry rule of thumb repeated often enough that most programs treat it as settled.

The claim did not follow basic manufacturing physics. Process tolerances on formed or machined features typically depend on feature size because the relative contribution of variation is larger at smaller scales. Rather than accept the blanket number, dimensional data was requested across a range of feature sizes and geometries — not just the program's nominal part.

The data showed a clear size dependence. On large features, the supplier achieved roughly half the stated blanket tolerance — they were underselling their capability. On small features, the actual achieved tolerance was roughly double the stated value — they were overselling it. The blanket number was approximately their average across the range, misleading in both directions and most misleading at the feature sizes that governed the design constraints. Writing the tolerance specification as a function of feature size — derived from the supplier's own measurement data — resolved three constraints that had appeared to be in conflict under the blanket specification simultaneously: the component fit the available form-factor envelope, cleared assembly placement requirements, and was physically sized to benefit peak operating temperature.

Acceptable implementations for the supplier capability data exchange include a supplier-collaboration object under product-data control when revision-locked handoffs are mandatory, a controlled exchange folder with revision-stamped attachments and a named review log for smaller programs, or a shared Cpk dashboard with defined refresh cadence. What matters is that the capability data — Cpk values, measurement methods, sample sizes, and control chart history — arrives under the same revision control as the design requirements.

Producibility as a continuous signal

Producibility is not a one-time pre-launch checkbox.

Track it through:

- first article findings,
- supplier drift patterns,
- field-return signals,
- design change impacts.

Without ongoing tracking, supplier drift is not visible until integration — after tooling is committed and the design is locked.

Field test: pick one supplier on your current program. Can you name the three critical characteristics you are tracking from them, and point to a controlled record with their latest measured values? If no, the supplier data link is not yet decision-grade.

Part V - Adoption and Scale

18. Rollout Sequence: Triage Then System

Friday, 11:12 a.m., customer presentation already on the projector, someone still writing "full rollout — week one" on the shared doc.

The program is already behind, with unresolved blockers accumulating every cycle.

Someone proposes a full OS rollout with new templates, role changes, and weekly governance meetings starting immediately.

That sequence usually collapses adoption before controls stabilize.

◎

START WITH TRIAGE, NOT BROAD TRANSFORMATION

When a program is unstable, first objective is to stop bleed rate:

- unresolved decision pileup,
- truth-state contradictions,

- late risk discovery,
- blocked dependencies with no owner.

Do not launch 20 controls at once.

Install the minimum loop that restores decision quality.

Triage phase (first 2-4 weeks)

Timebox triage and keep it narrow.

Core controls:

1. ownership map for top active decisions,
2. one-page truth on fixed cadence,
3. risk register with named owners and triggers,
4. gate output discipline (decision / owner / date / artifact delta).

If these four controls stay stable for two cycles, leaders can shift from firefighting to planned execution.

Entry/exit criteria

Define triage done criteria before starting:

- top decision backlog reduced below the threshold declared at triage start,
- one-page and source records aligned for two consecutive cycles,
- escalation path functioning with response SLA,
- no unresolved "which value is live" conflicts on critical threads.

Without exit criteria, triage becomes permanent emergency mode.

System phase (after triage)

Once stable, layer deeper controls:

- requirement lifecycle rigor,
- technical evidence loops,

- supplier capability discipline,
- training and cadence for durability.

Sequence matters: stabilize triage controls before layering system controls.

System controls without triage stability feel like overhead — teams comply on paper and work around them in practice.

Common rollout mistakes

- launching policy before proving value on one thread (no documented evidence package from a first-thread pilot in the rollout log),
- adding templates without owner behavior change (template completion rate rising while decision closure time unchanged),
- measuring activity count instead of decision quality (rollout metric is document count or meeting attendance, not closure rate or surprise rate),
- ignoring local context and forcing one-size sequence (single triage plan applied identically across programs with different distress patterns).

Each mistake lowers trust in the OS.

Adoption metric that matters

Track behavior change, not document count:

- decision closure time (timestamp delta between decision logged and decision closed in the decision log),
- decision carry rate (count of open decisions still unresolved at the start of the following review cycle),
- late surprise rate (count of risk register additions made after the last gate without prior flagging),

- one-page/record mismatch rate (comparison of one-page status to source record at weekly cadence),
- reopen rate on closed decisions (count of items returned to open status in the decision log),
- PM status request frequency between reviews (a sustained drop signals that cadence artifacts are trusted).

If these are improving, adoption is real.

Six programs, one rollout, eight weeks. Template completion hit 90% across all six by week six.

Decision closure time was unchanged. One program had started filling in the prior week's templates retroactively — not to improve the record, but to show the completion number. Another held weekly governance meetings with no decisions recorded. The rollout had produced activity, not adoption.

The effort stopped. Two programs — the ones with the worst unresolved decision backlogs — received the four-control minimum with predeclared exit criteria. No additional templates. No governance overhead. Four weeks later, both met the criteria.

The next two programs were not told to adopt. They asked. The first two were visible proof — review meetings shorter, decision backlogs shrinking, leads no longer chasing status. That visibility did more than any mandate.

Field test: name one program control from this OS that your organization installed in the last 90 days. Does it have a named DRI, a controlled record, and at least one decision that traced through it?

If not, it was installed in form but not in practice — and the next program will not use it either.

19. Training, Cadence, and Failure Recovery

On a Friday six weeks in, the rollout still looked healthy.

Then one lead change, one urgent supplier issue, and one skipped review cycle later, the old behavior was back — a program quarter of uncontrolled drift before the cadence recovered. The relapse analysis found three gaps: the skipped review had been covered by a handoff email, not a backup-trained lead; the lead change had no ownership-transfer protocol; the supplier issue had triggered an exception meeting that displaced the review cadence without a recovery date.

Three controls were added: a named backup DRI for each cadence slot, a thirty-day cadence audit after any lead change, and an explicit recovery trigger — if any scheduled review is displaced, the next one is held within five days regardless of program pressure. Six months later, a second lead change hit the same team. The cadence held.

Training is behavior transfer, not slide transfer

New people do not need the full theory first.

They need operational moves in context.

Onboarding minimum:

1. ownership map and closure expectations,
2. one-page truth mechanics and source links,
3. risk trigger/escalation behavior,
4. decision record standard.

Teach these on live program threads, not in abstract slides only.

Cadence is the enforcement mechanism

Without cadence, standards drift into preference.

Minimum cadence set:

- weekly decision/risk sync (updated decision log and risk register entries),
- fixed one-page refresh rhythm (published one-page status before each review),
- gate prep/check with artifact traceability (gate artifact delta from the record system),
- monthly retro on control-loop failures (playbook update with named control adjustments).

Cadence should be predictable and short enough to survive workload spikes.

Cadence is the mechanism that makes the artifact trustworthy enough to read between meetings. The engineering lead who has locked a working cadence has bought the program a habit that survives the next bad week.

Failure recovery without blame loops

When the system slips:

1. identify the control that failed (owner, record, gate, risk, status),
2. identify why it failed (capacity, ambiguity, missing authority, poor handoff),
3. restore minimum behavior on one active thread,
4. capture adjustment in playbook.

Do not frame recovery as "who failed process."

Frame it as "which control failed under what condition."

Metrics for real adoption

Activity metrics mislead when closure and mismatch outcomes stay flat.

The rollout chapter defines each metric and its artifact source. This chapter's job is reading the trend over consecutive cycles — a single-cycle improvement can be noise; a sustained two-cycle trend is the signal that behavior is actually changing, not just reported:

- decision closure cycle time — trending down, or plateaued after initial improvement?
- reopen rate on closed decisions — still dropping, or holding flat while closure rate rises?
- one-page vs source mismatch frequency — gone, or recurring on the same artifact types?
- late risk discovery rate — at zero after two cycles, or spiking around milestones?
- unresolved ownerless decision count — cleared each cycle, or accumulating?

If these improve for two reporting cycles, adoption is real.

If only template completion improves while outcome metrics stay flat, the rollout produced activity, not behavior change.

Turnover resilience

Systems fail at role transitions unless handover is explicit.

Require handover package for key roles:

- active decisions and owners,
- top risks and triggers,
- current one-page state,
- unresolved escalations,
- next gate commitments.

This turns personnel changes into manageable events.

Keep the bar practical

Overly rigid process breaks under load spikes.

Overly loose process breaks when ownership or data is ambiguous.

Durable cadence means:

- strict on ownership and traceability,
- flexible on meeting format and local workflow details.

◎

Eight weeks later, same program. The lead changed. The supplier issue hit. The review cycle slipped by one day, not three weeks — because the one-page had been published before Friday, the risk register named an owner on the supplier item, and the incoming lead read both before the first review. The handover package was on the shared drive: active decisions, top risks, current one-page, unresolved escalations, next gate commitments — the incoming lead was fully current within one review cycle.

Field test: name the last time a control loop on your program failed — a gate arrived without evidence, a DRI was absent, a cadence slipped. Is there a playbook entry that names the trigger, the recovery action, and the owner? If not, recovery is improvised every time, and the same failure will return.

20. What This OS Does Not Cover

Friday, 3:38 p.m., the gate room clears with nods instead of another circular fight.

The one-page status matches the risk register for the first time in a month. Someone mutters they might actually make the school pickup — nobody rolls their eyes, because the schedule window they guarded is still the one on the controlled plan.

◎

Honest scope is part of that quiet finish: say what this operating layer **is**, and what it **never** pretends to be, before anyone labels it compliance theater or a substitute for domain rigor.

It does not replace specialist depth or legal obligations.

Out of scope (explicitly)

This OS does **not** replace:

- domain safety lifecycles and safety case obligations,
- regulatory compliance frameworks and certification requirements,

- detailed design-history / quality-system requirements,
- specialist technical disciplines taught in dedicated texts,
- contract/legal governance obligations with customers or suppliers.

If your product sits in a regulated domain, those obligations remain primary and non-negotiable.

If you work under design controls, here is how the OS sits underneath

The OS does not replace your design history file, your safety case, or your audit obligations. It mounts underneath them. Use the OS layer to keep ownership, requirement truth, gate output, and one-page status legible to your engineering team while the regulated artifacts continue to be the authoritative record for the auditor.

Practical mount points:

- **Decision records** feed the rationale fields your design history file or equivalent already requires — write the rationale once, cite it twice.
- **Requirement lifecycle** and **physics-first requirements** sit upstream of your formal verification trace; the controlled hypothesis becomes the verified requirement once evidence closes.
- **Risk register and "done on paper"** is the engineering view; your formal safety analysis (FMEA, FTA, hazard analysis) is the regulatory view. The same closure evidence supports both, but the regulatory artifact is authoritative for the auditor.
- **One-page truth** is internal program control, not a regulated artifact. Do not confuse the two.

If a regulator or notified body asks for evidence, the OS does not produce it. Your formal lifecycle does. The OS makes sure your team is not running on fiction while that lifecycle runs.

What this OS is for

This OS helps teams run those obligations with less chaos:

- clearer ownership,
- cleaner requirement/revision truth,
- faster risk signal propagation,
- more credible gate and status decisions.

It is a coordination layer that links specialists, records, and decisions without replacing domain frameworks.

Where specialist depth must enter

You will still need specialists for:

- advanced tolerance and variation work,
- reliability/life and statistical confidence planning,
- safety and hazard analysis rigor,
- manufacturing capability and measurement systems depth.

Lead responsibility is to scope the question, own the decision linkage, and ensure outputs update program records.

Non-negotiables that apply everywhere

Regardless of domain, keep these habits:

1. one owner per decision path,
2. one source of truth for live values,
3. explicit decision records,
4. risk and status tied to source evidence,

5. clear escalation triggers.

These are portable across product classes and org structures.

Failure mode if boundaries are ignored

Two common errors:

- treating this OS as complete compliance (dangerous overreach),
- treating this OS as optional soft process (missed leverage).

The right stance is explicit: run this OS as a control layer under real domain obligations.

How the OS itself can degrade

Every operating system has failure modes of its own. Name them so you can spot the drift early and correct it as a system problem, not a personal one.

- **Process becomes bureaucracy.** When operating habits start collecting their own approvals, audits, and meetings *about* the habits, you have rebuilt the dysfunction one layer up. Fix: every operating artifact must trace to a decision the program would otherwise miss. If it does not, retire the artifact.
- **DRIs become bottlenecks.** A named owner becomes the only path, and decisions stall on one calendar. Fix: the DRI's job is to frame and record, not to be present at every conversation. Delegated decision frames are still controlled.
- **The source of truth goes stale.** The live record stops being live. Status meetings start running from copies. Fix: the weekly evidence update is the live record, not the slide. If the slide is the primary artifact, the live record has already failed.
- **Executives override the decision record.** A controlled decision is reversed in a hallway; no record is updated. Fix: hallway

reversals are decisions. If they are not recorded with the same fields as a formal decision, the OS is being undermined and the program is back on personalities.

- **Too many gates slow learning.** Gates exist to change controlled values. If every cross-functional checkpoint becomes a gate, the program loses iteration speed without gaining decision quality. Fix: prune gates that have no controlled value to change — a checkpoint without a controlled value to change is not a gate.

If two or more of these are visible at once, the OS is in maintenance debt. Fix it before the next program event — do not run the program on a broken OS. If the gap is too large to close immediately, roll back to the last chapter where the OS was working and restart implementation from there, one chapter at a time.

What the OS delivers

The OS does not remove uncertainty — hardware work stays cross-functional and time-constrained.

The promise is narrower and more useful:

- fewer repeated arguments,
- earlier truth on risk and dependencies,
- faster closure on real decisions,
- fewer expensive surprises caused by preventable process failure.

Field test: name one problem your program has right now. Is it a process gap — no record, no owner, no decision path — or a personnel gap? The OS addresses the first. If the answer is personnel, this book has nothing left to offer on that problem. Find the process gap and fix that first.

If your team can see the same truth, decide faster, and revise honestly, this OS is doing its job.

If that is true, shut the lid. Go home.

Artifact examples: inert vs. decision-grade

This page is a fast self-check, not a first teaching. The chapters explain why each field exists and what happens when it is missing; this page consolidates the four core artifact shapes so you can run a quick test against your own documents without re-reading a chapter.

The four artifacts below are the core surfaces the Hardware OS runs on. Every chapter that mentions "decision record," "requirement," "risk row," or "one-page status" is referring to these shapes.

◎

Requirement

The inert version has a value but gives the team nothing to test, no condition to bound it, and no way to know who approved the current number or why.

Field	Inert	Decision-grade
Statement	"System shall be robust"	"Leak rate shall be ≤ 0.5 sccm at 1.2 bar differential, ambient 0–40 °C, measured by Method M-04"
Owner	*(blank)*	Named engineer, week of last review
Evidence	*(blank)*	Link to test result or physics derivation that bounds the value
Acceptance method	"TBD"	Specific test method, sample size, pass criterion
Revision history	*(blank)*	Change log with rationale and approver for each revision

See Ch. 5 for revision discipline and the hygiene checklist.

◎

Decision record

The inert version records an outcome with no trail. Anyone who reads it six weeks later cannot reconstruct the trade, cannot tell who approved it, and cannot find what else changed as a result.

Field	Inert	Decision-grade
Decision	"Team aligned to update thermal limit"	"Revise thermal case temperature requirement from 75 °C to 78 °C, per first chamber test evidence"
Owner	TBD	Thermal lead
Rationale	*(blank)*	"First chamber test measured 78 °C steady-state; boundary condition error in thermal model confirmed"
Alternatives	*(blank)*	"Considered 80 °C with derating margin — rejected: supplier cannot hold derating at current geometry"
Downstream artifacts updated	*(blank)*	"Supplier package flagged for re-qualification, thermal risk row updated, Gate 0 one-page revised"
Approver / date	*(blank)*	Program lead, Day 9 after first chamber test

See Ch. 4 for DRI ownership and minimum decision record fields.

Risk row

The inert version looks like an active register. No one can trigger a decision from it, and “yellow” carries no evidence anyone can challenge or update.

Field	Inert	Decision-grade
Statement	"Thermal"	"Thermal margin insufficient at full-duty 40 °C ambient"
Owner	Engineering team	Thermal lead
Likelihood / Impact	Yellow	Medium (first chamber test showed 78 °C vs 75 °C limit) / High (gate hold, supplier re-qual)
Trigger	*(blank)*	Any chamber run exceeding 74 °C
Mitigation	In progress	Revise thermal case temperature requirement; re-qual supplier package by Gate 0
Decision date	*(blank)*	Gate 0
Evidence	*(blank)*	First chamber test, thermal model

See Ch. 6 for the full risk-row schema and gate decision logic.

One-page status

The inert version compresses uncertainty into confidence — an exec or PM reading it cannot tell whether the calm tone reflects real evidence or polished slides. When it is wrong, nobody finds out until it is too late to act cheaply.

Section	Inert	Decision-grade
Current state	"Program tracking to plan"	"Gate 0 on track; thermal risk open — thermal limit revision decision due Friday; supplier re-qual starts Monday"
Top risks	"Thermal risk tracking green"	"Thermal: medium/ high — first chamber test measured 78 °C vs 75 °C limit; owner: thermal lead; decision: Gate 0"
Decision log deltas	*(blank)*	"Thermal limit decision closed: requirement revised to 78 °C; supplier notified"
Critical dependencies	"Supplier on track"	"Cold-plate extrusion: slip of 8 working days; replanned to Week 22; critical path impact: none if Gate 0 holds"
Asks / escalations	"None"	"Need program lead approval on revised supplier qualification plan by Thursday"

See Ch. 6 for the minimum one-page schema and source-record traceability requirement.

Substitute your program's artifacts; the field shapes hold.

www.ingramcontent.com/pod-product-compliance
Lightning Source LLC
LaVergne TN
LVHW022226200926
848481LV00006B/2140